MEDIA AND NEW CAPITALISM
IN THE DIGITAL AGE

MEDIA AND NEW CAPITALISM IN THE DIGITAL AGE

THE SPIRIT OF NETWORKS

Eran Fisher

First published in hardcover in 2010 by PALGRAVE MACMILLAN®
in the United States—a division of St. Martin's Press LLC, 175 Fifth
Avenue, New York, NY 10010.

Where this book is distributed in the UK, Europe and the rest of
the world, this is by Palgrave Macmillan, a division of Macmillan
Publishers Limited, registered in England, company number 785998,
of Houndmills, Basingstoke, Hampshire RG21 6XS.

Palgrave Macmillan is the global academic imprint of the above
companies and has companies and representatives throughout the
world.

Palgrave® and Macmillan® are registered trademarks in the United
States, the United Kingdom, Europe and other countries.

ISBN: 978-1-137-31081-1

The Library of Congress has cataloged the hardcover edition as
follows:

Fisher, Eran.
    Media and new capitalism in the digital age : the spirit of networks /
by Eran Fisher.
        p. cm.
    Includes bibliographical references and index.
    ISBN 978-0-230-61607-3
    1. Information technology—Social aspects. 2. Information
technology—Economic aspects. 3. Technological innovations—Social
aspects. 4. Technological innovations—Economic aspects.
5. Capitalism—Social aspects. I. Title.

    HM851.F575 2010
    303.48'33—dc22   2009023181

A catalogue record of the book is available from the British Library.

Design by Scribe Inc.

First PALGRAVE MACMILLAN paperback edition: September 2013

10–9–8–7–6–5–4–3–2–1

To my mother and father, with love and gratitude

Contents

ILLUSTRATIONS

FIGURES

TABLES

ACKNOWLEDGMENTS

Many people have helped make this endeavor possible, enjoyable, and worthwhile, and it is a pleasure for me to thank them for their friendship and support.

First among them is Efrat Eizenberg, who has been my companion at every twist and turn and whose limitless friendship and support has been my main source of comfort.

Thanks to my family for their unwavering support and encouragement. My first and foremost teachers: my wonderful sister Orit Feinstein, and my mom and dad, Rina and Karol Fisher, for whom I dedicate the book. Thanks also to Lior, Noa, and Nadav Feinstein; Nilli and Hezi Eizenberg; Elad Eizenberg and Nicole Sears; and Ahuva, Tali, Moshe, and Keren Plosker. Lastly I want to acknowledge the new edition to my life, my daughter Roya.

Thanks to my friends Iria Candela, Isabel Cuervo, Elinor Dahan, Nada Gligorov, Sherrie Glass, Tammy Golan, Iris Gur-Arye, Roger Hart, Tsai-Shiu Hsieh, Haddassa and Marion Kosak, Racheli and Amir Kraus, Nicholas Leclercq, Oren Orevi, Ilya Ram, Solen Sanli, Nava Schreiber, Barry Spunt, Moran and Ari Tenenbaum, Zeynep Turan, Didem Yilmaz, Dvora Zak, and Dana Zelinger. Thanks to my friends at the New School, Yuri Contreras and Yoav Mehozay, who have read and commented on parts of this work and were always there to offer encouragement and advice. And to my dear friends Guy Abutbul and Breixo Viejo—with humor and wisdom, they always managed to calm my fears and concerns without ever taking them lightly.

This book began as a doctorate dissertation at the Department of Sociology at the New School for Social Research in New York. At the New School for Social Research, I would like to thank Jeffrey Goldfarb for his support of my project. Thanks to Paolo Carpignano, my dissertation advisor, who has been immensely supportive and encouraging and provided invaluable professional feedback. Special thanks to Stanley Aronowitz of the Graduate Center of the City University of New York for willing to be part of the dissertation committee. It was a great privilege to have him contribute his academic expertise and exhilarating personality.

The publication of this book has been supported by a research grant from the Department of Behavioral Science, The College of Management-Academic Studies in Israel. I want to thank the department and especially Amir Ben-Porat and Tova Bensky for their generous support. I would also like to thank my editors at Palgrave, Julia Cohen and Samantha Hasey, for their professional support and personal courtesy.

I would like to thank my teacher and friend, Yishai Tobin of Ben-Gurion University in Israel, who facilitated my first steps as an aspiring scholar over a decade ago and has not withheld his generous support and care ever since.

Lastly, I want to thank Uri Ram, of Ben-Gurion University and The New School, whose influence and contribution to my scholarly upbringing and to this work in particular is immeasurable. I am deeply thankful for his generosity, thoughtfulness, and sincerity and for his friendship.

TECHNOLOGY DISCOURSE AND CAPITALIST LEGITIMATION

TECHNOLOGY AND SOCIETY

SOCIETY IN THE LAST FOUR DECADES HAS WITNESSED TWO extraordinary transformations in advanced capitalism. One is the transformation of capitalism from Fordism to post-Fordism, involving changes not only in the regime of accumulation—that is, in how production is carried out, where, and by whom—but also in the mode of social regulation, entailing a whole set of political arrangements and cultural practices. This has been a deep social transformation: globalization, the "new economy," Google, outsourcing, "just-in-time" production, the rise of India—these are just few of the new keywords in the lexicon of the new capitalism. The other transformation was the emergence of network technology (or information and communication technology) and its integration into virtually every sphere of life. This has been nothing short of a technological revolution: indeed, many of the keywords in the lexicon of the new technology parallel those of the new capitalism.

The question then arises: what is the nature of this close affinity between new capitalism and new technology? The answer seems straightforward: a new technology enables a new society. Globalization is carried over networks of communication. The "new economy" is essentially all about new businesses and consumer products centered on the value of information. Google epitomizes such a business model. Outsourcing and "just-in-time" production are hard to imagine as viable economic practices without network technology. India owes its rise as a capitalist miracle to customer-service call centers in Bangalore and to the surge in the number of software engineers and global hubs for the high-tech industry.

This outlook reflects a prevailing assumption: technology makes society. For example, such a view underlies much of the writing of prominent *New York Times* columnist Thomas Friedman, whose image of the world today (i.e., flat and decentralized) follows the imagery of the architecture of network technology. Freidman not only employs network technology as a conceptual framework but also sees the unique features of such technology percolating into social dynamics, claiming, for example, that "the Internet offers the closest thing to a perfectly competitive market in the world today" (Friedman 2000, 81).

According to this view, to talk about digital technology is both *transparent* and *instructive*. It is transparent because the discourse on technology is merely a reflection of the realities of digital technology (to say, for example, that network technology allows flexibility and interactivity is to merely state the obvious). It is instructive because the discourse ultimately describes the operation of society (e.g., the flexibility of digital technology creates a flexible economy). However, I argue here, the discourse on technology is not a transparent vignette on reality but rather a particular outlook on it and is therefore worthy of analysis in its own right. Moreover, technology not only constitutes the material foundations of modern societies but also functions as its legitimation. The prevailing outlook regarding the relationship between technology and society is partial and, as such, ideological as well. It is the ideological nature of these relationships between (a new network) technology and (a new, post-Fordist) capitalism that I want to make the center of my investigation.

HISTORICAL TRANSFORMATIONS IN TECHNOLOGY DISCOURSE

This book offers an inquiry into the contemporary discourse on network technology in advanced capitalist societies—in short, the *digital discourse*—that is located between a new capitalism and a new technology. I will argue that the sociopolitical constellation of our current network society is the result not only of economic and technological transformations but also of *ideological* transformations. That technology functions ideologically in modern societies is evident at almost all social levels, from the cultural realm where technological gadgets and engineering feats are celebrated and seen as pinnacles of the human spirit, to the political realm where instrumental reason reigns and technological means overwhelm substantive rationality.

But beyond this general critique of a technologistic outlook are the *particular* forms of technology discourse that are historically and sociologically contingent. Hence, while technology discourse can be said to function ideologically, its particular role and content are dependent upon specific social constellations and are variably integrated into the political culture of modern societies. At the center of this book is the argument that the role of technology as a legitimation discourse has gone through radical transformations concurrent with the transformations of capitalism in the last few decades. During Fordism, technology discourse legitimated the interventionist welfare state, the central planning in businesses and the economy, the hierarchized corporation, and the tenured worker. However, during our contemporary, post-Fordist society, technology discourse legitimates instead the withdrawal of the state from markets, the globalization of the economy, the dehierarchization and decentralization of businesses, and the flexibilization of production and the labor process. This new legitimation discourse of technology marks a transformation in the spirit of capitalism (Weber 1958) to use Max Weber's famous expression, from its industrial phase, which emphasizes the capacity of capitalism to bring about social emancipation by alleviating exploitation, to its postindustrial spirit-of-networks phase, which focuses on the capacity of capitalism to enhance individual emancipation by alleviating alienation.

Bringing together ideas from the Frankfurt School's critique of instrumental reason and technocratic legitimation with Boltanski and Chiapello's recent revival of a critical economic sociology, I argue that the legitimation of capitalism and its transformations are closely intertwined with the critique of capitalism in general and technology in particular. That is, that the shift from a Fordist to post-Fordist society has also entailed a shift in the legitimation discourse of technology. During Fordism, technology served to legitimate a regime of accumulation that favored a response to the social critique of capitalism while downplaying, and even ignoring, the humanist critique of capitalism. Industrial technology, the assembly line, the bureaucratic corporation, the statist regulation of the economy, and the provision of welfare were all conceived in technology discourse during Fordism as techniques and technologies that respond to concerns put forth by the social critique, such as social security, stability, and equality—that is, those geared toward mitigating exploitation.

Hence the critique of Fordism targeted precisely the oppressive nature of the administered state and the bureaucratic corporation, the loss of personal authenticity, and the de-eroticization of the productive process, that is, the harmful ramifications of Fordism and industrial machines and techniques in terms of alienation. One of the brilliant articulations of that type of critique of industrial capitalism appears in Charlie Chaplin's *Modern Times* (1936). The poor tramp finds it hard to keep up with the pace of work, which is dictated by the assembly line, gets hit by a feeding machine, which is supposed to make his work more efficient by cutting out his lunch breaks, and is ultimately swallowed up by the machine and gets integrated into the clockwork that propels the industry. According to this prominent critique, the rationalization of society by technological means may have oiled the wheels of capitalism, but that came at the price of hindering personal freedom and creativity, alienating workers, and, more generally, creating a centralized, bureaucratized society.

With post-Fordism and with the development and integration of network technology into virtually all spheres of life, technology discourse has come to legitimate a new social regime that responds to the humanist critique of capitalism, while it downplays the social critique. Network technology (primarily the Internet), the lean and decentralized corporation, and flexible modes of employment and production are all conceived in the discourse on network technology during post-Fordism (i.e., the digital discourse) as technologies that respond to concerns put forth by the humanist critique, such as individual empowerment, authenticity, and creativity—that is, those geared toward mitigating alienation.

The ramifications of these transformations go beyond the relatively limited scope of a technological sphere, or even that of capitalism. Rather, they should be understood within the wider scope of the constellation of power and the political culture of the post–World War years of Western capitalist societies. While the discourse on industrial technology during Fordism was used to legitimize the social compact between capital, labor, and the state, the discourse on network technology during post-Fordism legitimizes precisely the decomposition of this compact and the constitution of its alternative: privatized relations within the context of a global market and civil society. The two political cultures of Fordism and post-Fordism should also be distinguished by the very different political significance they assign to technology in society. While Fordism extolled the power of technology in the name of social equality and stability as a public and political project, post-Fordism extols the power of technology in the name of individual authenticity and liberation as a private and apolitical (or postpolitical) enterprise.

The articulation and substantiation of these arguments is the task of this book. Part I lays out in further detail the theoretical, sociological, and empirical foundations. Chapter 1 locates the transformations in technology discourse within the contours of the legitimation of capitalism in general and lays out the theoretical foundations for the study of technology discourse as a legitimation discourse. It then presents the main transformations in capitalism in the last few decades along two lines: a shift in the mode of accumulation from the Fordist phase of capitalism to its flexible, post-Fordist phase and a shift in the mode of regulation encompassing a new constellation of power between three central social actors: labor, capital, and the state. The chapter concludes by tying the arguments together around the notion of the "spirit of networks," which characterizes our contemporary brand of techno-capitalism.

The book offers not only a theoretical framework to account for the new spirit of networks but also a detailed empirical account of one of the most crystallized voices of the digital discourse—*Wired* magazine. Chapter 2 discusses the centrality of the digital discourse in contemporary public discourse and establishes the centrality of *Wired* in the emerging digital discourse in contemporary political culture and, hence, its adequacy as a case study of the digital discourse.

FOUR LOCI OF NETWORK CAPITALISM

The digital discourse is crystallized around four key transformative sites at the intersection of network technology and the new, post-Fordist capitalism: the market, the world of work, production, and the conception of humans. The analysis of these sites is presented in Part II of the book. Chapter 3 examines the discourse at the intersection of network technology and the market. According to the digital discourse, as markets are integrated into network technology they become more frictionless and rational, demand almost no external political regulation, become more chaotic and unstable, and, in turn, require nodes—both individuals and companies—to be flexible and adaptive to an ever-changing environment. These narratives describing a network market are understood to be an ideological articulation of the transformation from the embedded liberalism of the Fordist era to the neoliberalism of the post-Fordist era. Within a technologistic framework, this discourse legitimates the trend whereby markets become more autonomous and disembedded from social regulation and are seen as an ideal mechanism for the administration of social reproduction.

Through a comparison of the digital discourse (centered on technological arguments) and the explicit neoliberal discourse (centered on economic arguments), Chapter 3 deconstructs the ideological foundations of the notion of *network* markets in the digital discourse. In the digital discourse, the notion of networks is rooted and constructed within the paradigm of markets—that is, as a template to understand not only economic markets but also society *as* a market. Set within the theoretical framework of this book, this chapter argues that the discourse on network markets legitimizes and depoliticizes the withdrawal of the Keynesian welfare state and the emergence of a new regime of neoliberalism. The chapter also serves as a general introduction to the key words that dominate the digital discourse as a whole, such as network rationality, spontaneous order, chaos, decentralization, and dehierarchization.

Markets are not the only systems transformed by network technology: the whole system of production changes as well. The analysis of the digital discourse on these transformations is presented in the next two chapters. Chapter 4 examines the digital discourse at the intersection of network technology and the world of work. According to the digital discourse, as the traditional world of work is integrated into network technology, the boundaries between work life and personal life become indistinguishable; work space and work time are intermingled with their private, personal counterparts. These novelties allow workers to bring their personal, lifeworld qualities of creativity, intimate relationships, and deep personal engagement to bear on their work activities and reeroticize the disenchanted world of (industrial) work. They also render the workplace flat—that is, decentralized and dehierarchized—and eradicate hierarchies between workers and managers and owners. What results is a regime of meritocracy and professionalism based on one's proficiency with network technology. These narratives of network work signal a radical break from the narratives of work alienation in the bureaucratized organization. Network technology is situated at the pivotal axis of rendering work—the work process, work relations, and the workplace—more humane and more liberating for the worker.

At the macro level, these narratives also signal a shift from a Fordist discourse of *class* to a post-Fordist discourse of *networks*. In the discourse of networks, individuals are construed as autonomous nodes and defined by their connections to other nodes in the network; the social is seen as a flat, decentralized sphere of ever-flowing, multiple, and ad hoc assemblages. This is a stark contrast to the Fordist conception of the social sphere as consisting of a hierarchized, stable, and category-defined arena. While the discourse of class stresses structural power relations, the discourse of networks is devoid of such forms of power (except, admittedly, the lack of power that results from being cut off from the network, i.e., being at the dark side of the digital divide) and instead associates power with the characteristics of autonomous nodes (i.e., power resulting from ingenuity and entrepreneurship). In the same vein, the central mode of social action in the discourse on networks is that of cooperation rather than struggle or competition, which characterizes the discourse on class. Hence, the notion of network work and its grounding in a technological reality allow for the substitution of a hierarchical, competitive, and antagonistic model of class by a dehierarchized, cooperative, agreeable, and inherently inclusive model of networks (at least in the long run, once the digital divide is bridged).

While Chapter 4 focuses on the ramifications of network technology for the traditional (or industrial) world of work, mostly within the confines of the "company," or the "workplace," Chapter 5 focuses on new forms of work and production brought about by network technology and taking place outside of these traditional sites. Chapter 5 examines the digital discourse at the intersection of network technology and the process of production. The integration of production into network technology results in the blurring of the boundaries between companies and the network and between producers and consumers. The network becomes the prime locus and axial coordinator of production, replacing the company, the workplace, and the assembly line. It facilitates new productive modes that are predicated on the ability to harness the working power of autonomous and dispersed nodes into the productive process.

From the point of view of individuals, these new productive modes allow more people to engage more meaningfully with—and to bring their skills, talents, and passions to bear more fully upon—the productive process. The productive process becomes more democratic and collaborative and is geared more fully toward personal fulfillment of the lifeworld rather than toward the fulfillment of system's ends. Moreover, network production makes possible the perfect *fusion* of the needs of personal emancipation with the system's needs of capitalism. These narratives amount to a discourse where the very notions of work, workers, and certainly the working class are undermined and in fact eliminated. Work is reconceptualized as an eroticized, playful activity of production and consumption, involving creativity, deep engagement, interactivity, and interpersonal communication. Likewise, the category of worker is substituted by a new category of the *prosumer*: an individual and autonomous unit of production, consumption, and entrepreneurship. Therefore, the discourse on network production legitimizes a shift to a post-Fordist organization of labor and production. The sites where the social organization of work during Fordism were contained and anchored—the company, the union, the professional, and the cadre—are rendered obsolete with the emergence of the prosumer. At the same time that work becomes more meaningful and humane, allowing greater outlet for personal potential and harnessing amateurish skills and leisure time into social reproduction, it also becomes more privatized and individualized, shifting more risks from capital to labor and dismantling the social buffer zone offered during Fordism—a buffer zone that favored social equity and personal security over the development of individual potential. The digital discourse presents network production as making the production process both more democratic and more engaging and, at the same time, undermining the institutional arrangements that made those processes more stable and protective. As such, the digital discourse is understood here to legitimate more flexible and precarious employment arrangements, which have become pervasive in post-Fordist capitalism.

The transformations in the discourse on the network market, network work, and network production lead to and are dependent on a fundamental shift in the perception of the individual, that is, how humans are perceived and understood in the context of a technological revolution. The perception of humans is intimately linked with the dominant technology with which humans engage (either in production or communication) and is therefore historically contingent and changes concurrent with technological transformations. Chapter 6, therefore, examines the digital discourse at the intersection of network technology and the individual. In this discourse, humans are understood to share the same essence and characteristics of network technology: both are seen as technologies that store and process information, and both are deemed to be flexible and indeterminate. The commensurability of network technology and humans allows them to interact more tightly and even converge. The integration of humans into network technology results in a new conception of the individual as network and cyborg. The blurring of the boundaries between humans and network technology allows for a more meaningful, emancipatory, and natural interaction of humans with the technology and unleashes the emancipatory potential of humans: their intellectual, psychological, and communicative abilities. The network human is seen as an emancipated version of the previous industrial human: the network human is now able to engage with the technology and harness his mental skills for network production. This contrasts with the industrial human, who was dominated by technology and was able to harness merely her physical and manual skills for industrial production.

These narratives are weaving a spirit of networks for post-Fordist society. Humans are reconceptualized to fit within the new mode of production: contrary to the Fordist human characterized by spatial and temporal bodily presence and physicality, the post-Fordist human is characterized by virtuality and disembodiment. The post-Fordist human—her body, mind, and identity—are informational, hence flexible and multiple. Network technology is thus seen as a response to the humanist critique regarding authenticity; network technology complements the network human.

THE POLITICAL CULTURE OF NETWORK CAPITALISM

Together, the narratives of the digital discourse explored in Part II of the book amount, I contend, to a new spirit of networks that legitimizes the new constellation of power that characterizes post-Fordist society: the withdrawal of the state from the planning, management, and regulation of the economy and from its welfare obligations; the move from a national, protective economy to a globalized, deregulated, and unitary market; the privatization of work; the eradication of work and the working class as viable social categories; the disembeddedness (a term coined by Karl Polanyi) of markets from society or, more specifically, the insulation of the economy from the democratic political process; new employment schemes that are more precarious and flexible; a new, decentralized, dehierarchized disciplinary regime of production; a shift to a placid and technocratic political discourse of a classless society, devoid of antagonism and contradictions; the increased flexibility of working life and of labor processes; and the depolitization of economic affairs as technical, technological, and instrumental.

Part III of the book is comprised of two concluding discussions that articulate two broad sociological and theoretical dimensions of the digital discourse. Chapter 7 defines and explains the *technologistic* underpinnings of the discourse on network technology. It presents the new cosmology of network technology that prevails in the digital discourse and asks—in the context of the transition from Fordism to post-Fordism—what is the social meaning of a discourse that revolves on technology as its axis of understanding and presentation?

This chapter not only adds another floor to the structure of the digital discourse constructed thus far but also runs across the previous floors like a beam and therefore brings us closer to the conclusion of this book. According to the digital discourse, network technology constitutes the teleological climax of the history of not only technological progress but also information, binarism, and indeed the universe. The new realities of network markets, network work, network production, and network humans discussed in Part II, which are in turn predicated on the realities of network technology—flexibility, adaptability, dehierarchization, decentralization, self-regulation, individualization, communication, spontaneous order, and so forth, represent a technological (and, in turn, social) revolution in the original sense of the word: a return to the very essence of nature and the universe. This is portrayed in stark contrast to industrial technology, which represented a breaking away from nature.

The spirit of networks is not simply a discourse about the overcoming of the discontents of industrial technology and the industrial condition (alienation, centralization, environmental degradation, and so forth) by network technology. More than that, it is a discourse about our late capitalist civilization, our human condition, our society representing the end of history (Fukuyama 1992). This staple of contemporary political culture in advanced capitalist societies—epitomized in the maxim (at once both postideological and programmatic) "There is no alternative," famously coined by Margaret Thatcher—receives technological clothing in the form of network cosmology. Network cosmology provides a technological validation for the crisis of ideological critique in the political culture of capitalist societies, that is, the widespread belief that such critique is futile and that we are living in a postideological time defined in fact by technology, not ideology.

Chapter 8 recapitulates the central narratives of the digital discourse about contemporary society, demonstrates how these narratives crystallize into a coherent "spirit of networks," and situates this spirit within the broader context of contemporary political culture, specifically the discursive shifts that have been occurring in the last four decades—shifts from distribution to recognition, from equality to diversity, from political economy to identity politics, and from the social critique to the humanist critique of capitalism.

The digital discourse, I argue, represents a radical transformation in the techno-political culture between modernity and postmodernity. While the political metanarrative of modernity and, more specifically, the modern welfare state has been that of *promise*, in postmodernity political action has been reconstrued as a necessity and *unavoidable adjustment*. No longer is politics the management of social projects; now it is a degovernmentalization of the state. In this context the digital discourse offers a new technological metanarrative of postmodernity, which stands in stark contrast to that of the modernist era. During modernity, the discourse on technology delivered a utopia of planning, control, and proactivity; it situated technology as the axis of a sociopolitical project. The new metanarrative of technology during postmodernity is one of technology as a tool for flexibility, adaptability, and reactivity. In addition, this chapter offers concluding discussions regarding contemporary political culture and the role that the discourse on technology plays in it.

THE NEW SPIRIT OF TECHNO-CAPITALISM

To conclude, at the center of this book is the argument that network technology has helped bring about a social revolution not only by its instrumental, functional, and material force but also by its discursive force. The discourse on network technology has been a centerpiece in the legitimation of the transformation of advanced capitalist societies from Fordism to post-Fordism. The extraordinary social, economic, political, cultural, and personal ramifications of this social revolution are given in this discourse a technologically centered framework that renders those transformations neutral, natural, inevitable, and benevolent. In the digital discourse, network technology is constructed as the axial factor that allows for the overcoming of the core problems associated with industrial,

Fordist society: the bureaucratic organization of work, the heavy-handed interventionist state, the diminishing levels of economic growth, the pre-eminence of collective interests over personal freedom, the drudgery of industrial work and its incompatibility with humans' diverse capabilities and needs, and the political culture of competing ideological viewpoints. The integration of network technology into markets and into the process of production allows for the materialization of a friction-free, global, self-regulated market (overcoming national markets hindered by interventionist states); the mobilization of information and networks as new forces of production (overcoming limits to economic growth); the dehierarchization and decentralization of the workplace and the work process (overcoming disenchanting bureaucratic organization and class conflict); and the rendering of work as a creative, cooperative, and personally satisfying process (overcoming alienation from the labor process). Finally, this techno-social constellation of a strong market, a weak state, an empowered individual, and a privatized social sphere is presented as the pinnacle of human progress and as an apolitical situation that has no political alternatives and is hence heralded as the end of history.

The spirit of networks is the new spirit of contemporary techno-capitalism, offering a technologistic framework to the structural transformations of capitalism from Fordism to post-Fordism. This hegemonic discourse offers both a *critique* of Fordist society and its techno-political culture of social empowerment and the overcoming of exploitation, and an *affirmation and legitimation* of post-Fordist society and its techno-political culture of individual empowerment and the overcoming of alienation. But as this book shows, while contemporary digital discourse is explicitly critical and emancipatory, it is in fact implicitly the hegemonic discourse of contemporary network capitalism *par excellence*, legitimating a new trade-off between social emancipation and personal emancipation.

PART I

CHAPTER 1

CAPITALISM, TECHNOLOGY, AND THE DIGITAL DISCOURSE

EXPRESSIONS OF WONDER AND AWE IN THE FACE OF new technology dominate public discourse and are indeed hard to contain. From new media technology to the Internet, from Google to global positioning systems (GPS) and cellphones—the overwhelming feeling of novelty and ingenuity embodied in these technologies—many of which are experienced firsthand by millions of individuals around the world—can easily slide toward what Vincent Mosco calls *The Digital Sublime* (2004), a fascination with (if not fetishization of) technology and its tremendous impact on our everyday life experience. The digital discourse is indeed precisely that body of knowledge that epitomizes this contemporary awe and the feeling that network technology changes everything, remaking society in its own image. But, as I have already pointed out in the Introduction, notwithstanding the tremendous ramifications of network technology on contemporary society, technology is not only the material basis of society but also its ideological foundation. Technology discourse is not a transparent vignette on reality but rather a direct influence on the construction of reality and is therefore worthy of analysis in its own right.

THE DISCOURSE ON TECHNOLOGY

TECHNOLOGY AND "POST" THEORIES

I understand the discourse on technology as a cognitive map, a structure of feelings, and an episteme; that is, a body of knowledge that is inextricably intertwined with technological reality, social structures, and everyday practices (Foucault 1994; Jameson 1991; Williams 1978). The digital discourse is a public discourse that situates network technology at the center of an emancipatory social transformation. This thesis has been crystallized within a few theoretical frameworks, most notably

postindustrialism, postmodernism, and posthumanism, and is the bedrock of media studies and, more recently, cyberstudies (Webster 2005).
For the postindustrialists, the determining role of knowledge, information, and technology in the productive process and the corollary decline
of the working class and rise of a professional-technocratic elite—bent on
rational planning and affiliated with neither capitalists nor workers—also
implies the substitution of a rational, technocratic political sphere for the
ideological politics of class struggle and the strengthening of civil society
(Bell 1999; Machlup 1962; Porat and Rubin 1977; Touraine 1971).

For the postmodernists, the dissociation between signifier and signi-
fied and the constitution of an "empire of signs" (Barthes 1982) or a
"hyperreality" (Baudrillard 1983) "*media*-ted" (Lash 2002) by network
technology is at the heart of a radical social break and a move to a new
"mode of information" (Poster 1990). Like the postindustrialists, they,
too, uphold the emancipatory potential of this informational transformation: liberation from grand narratives, from essentialist and authoritarian bodies of knowledge about the world, from metaphysical ontologies,
and from the determination of signs (Baudrillard 1975, 1981; Foucault
2002; Lyotard 1984). In the same vein, posthumanists are excited about
the constitution of new subjects defined by new advances in information
and communication technology. The informationalization of the body
and the networking of identities allow for a more negotiated and indeterminate construction of one's identity, for the overcoming of essentialist
categories not just in discourse but in practice, and for more degrees of
freedom in choosing one's identity, which in turn opens up new opportunities for equality, especially for previously underprivileged subjectivities,
such as women (Gaggi 2003; Haraway 1991, 1997; Hayles 1997, 1999;
Turkle 1997). To sum up, postindustrialists, postmodernists, and posthumanists all locate network technology at the heart of a radical break in
social life that helps transcend the Achilles' heel of industrialism, modernism, and humanism, respectively.

THREE APPROACHES TO THE RELATIONS BETWEEN TECHNOLOGY AND SOCIETY

The approach I take in this book can be seen as a critique of these three
prominent theoretical formulations inasmuch as I ask to shift our attention from the social effects of the materiality of technology to the social
effects of the ideological facet of technology and investigate technology
discourse. To understand the status of "technology discourse" as a sociological object of study, we need to locate this mode of analysis within
the broader field of the social study of technology. Three theoretical
approaches have been proposed to account for the relationship between

technology and society. The most prevalent approach is that technology shapes society, thus framing the inquiry in terms of the effects of technology on society. Underlying this "technologistic" approach (Robins and Webster 1999) are three assumptions: *neutrality*—technology has a history of its own, its development stems from its internal dynamics, and it is therefore an asocial force, external to social power struggles (Bijker 1995; Robins and Webster 1985; Feenberg 1995); *inevitability*—technology determines the shape of society, reconfiguring it in accordance with its internal workings (Winner 1977; Smith and Marx 1994; Feenberg 1991); and *benevolence*—technological progress is human progress, as it is positive in and of itself (Robins and Webster 1999; Postman 1993).

The uncovering and overcoming of these problematic assumptions (often denounced as "technological determinism") have underpinned critical approaches to technology, which suggest that society shapes technology. Such approaches seek to introduce social coordinates into the analysis of technology construction, dissemination, and use.[1] These critical approaches share "an insistence that the 'black-box' of technology must be opened, to allow the socio-economic patterns embedded in both the content of technologies and the processes of innovation to be exposed and analyzed" (Williams and Edge 1996). Likewise, late Marxist analysis has been particularly fruitful in uncovering the extent to which technologies of production have been means in the arsenal of class struggle, rather than the mere implementation of universal instrumental rationality (Braverman 1974; Noble 1984; Aronowitz and DiFazio 1994; Huws 2003).

Despite the contrast between these two approaches, both share an engagement with technology as an instrument. The third intervention into the relationship between technology and society—and the one that informs this book—focuses on technology as discourse—cultural, social, political, and ideological. According to this approach, the discourse on technology is not simply a reflection of the centrality of technology in the operation of modern societies; rather, it plays a constitutive role in their operation and enables precisely that centrality. Some see technology discourse as a form of "projection" (Heffernan 2000) of social realities or a "technological vision" (Sturken and Thomas 2004) through which transformations of a political, economic, and social nature are filtered. Thus, for example, popular science fiction of the late twentieth century has articulated the new, post-Fordist relationship of production between capital and labor (Heffernan 2000), just as the emerging technology of the mechanical clock in seventeenth-century Europe played a central role as a metaphor and cognitive framework to deliberate the ideological debate between authoritarian and liberal conceptions of political order (Mayr 1986).

A stronger version of that approach sees technology discourse as assuming a more active role in the construction of reality, a reality that it presumes merely to describe. Here, technology discourse is seen as central to shaping the political, cultural, and social Zeitgeist. Nye, for example, argues that the "American technological sublime"—the reverence of technology in and of itself—has become at the turn of the twentieth century "one of America's central 'ideas about itself'" (Nye 1994, xiv), functioning to reaffirm and bind multicultural individuals into a society by celebrating and admiring technological achievements in the public sphere. Technology discourse played an equally central role in the construction of nationalistic culture in Germany's Third Reich, where technological rationality was co-opted by irrational politics to create a political culture of "reactionary modernism" (Herf 1984), with technology helping to define the emerging German national fascist identity. Lastly, the emergence of the discourse on "the human motor" (Rabinbach 1992) during the nineteenth century was interwoven with scientific and technological transformations: the scientific "discovery" of the laws of energy was central to a revolution in the perception of humans as containers of energy, leading to new practices, such as Taylorism, which in effect rendered the human body a motor and mobilized it to the mechanical process of mass production.

The discourse on technology is anchored in the material and functional aspects of technology. It is precisely this anchoring in objective reality that renders it a discourse in the strong Foucauldian sense: a quasi-scientific set of truths, which are beyond dispute and hence serve as assumptions for further reflection on reality and engagement with it. It also points to the fact that a discourse is not an ideal construction—imposed on reality in order to veil it—but a lens through which reality is understood and acted upon. Such an approach, I believe, allows us to bypass the deadlock of causality (does the discourse on technology create a certain social reality or vice-versa?; does technology drive history?; and so forth; Smith and Marx 1994; Castells 1996, 5) and points to the dialectic relationship between the discourse on technology and social practices that are part of a new social totality (Rabinbach 1992, 52; Kern 1983).

THE DISCOURSE ON TECHNOLOGY AS LEGITIMATION DISCOURSE

The strongest version of this approach to the social study of technology sees technology discourse as a particular outlook, an ideology. According to this view—developed particularly by members of the Frankfurt School as part of a broader critique of instrumental reason—with modernity, and specifically with the harnessing of science and technology for the needs of capitalism and the modern state, technology discourse has come to play a central role in the legitimation of a techno-political order, that is, a political order that is legitimated by technology and techniques (see Horkheimer and Adorno, 1976). In this political context technology becomes an unquestionable "good," a "religion" (Noble 1999), and a "myth" (Robins and Webster 1999, 151; Mosco 2004), which suggests that virtually any social problem is subject to a technical or technological fix (Aronowitz 1994; Segal 1985). It entails the substitution of technical and technological discussions, with their emphasis on instrumental rationality, for political debate based on communicative rationality and aimed at arriving at substantive rationality (Marcuse 1991; Habermas 1970; Feenberg 1991; Pippin 1995; Borgmann 1984b). The idea (indeed, the ideal) of technology gains in itself a constitutive role in society, becoming the point of departure and the yardstick against which other realms of society are examined and worked out; technology becomes "both means and the ends, the instrument of progress but also its fulfillment" (Robins and Webster 1999, 151). In sum, with modernity, "technology has become not just the material basis for society but in a real sense its social and ideological model as well" (Segal 1994, 3).

Furthermore, technology functions as an "ideological tool that mystif[ies] mechanisms of power and domination" (Best and Kellner 2000) and is "one of the major sources of public power in modern societies," commonly used as an alibi "to justify what are in reality relations of force" (Feenberg 1995). One of the most eloquent theoretizations of the ideological functions of technology discourse is offered by Jürgen Habermas (1970). Habermas points to the substitution of technical and technological discussions, with their emphasis on instrumental rationality, for political debate, which is based on communicative action and aimed at substantive rationality (see also Marcuse 1991; Feenberg 1991, 1996; Pippin 1995). In his seminal article "Technology and Science as 'Ideology,'" Habermas (1970) lays out a history of capitalist legitimation whereby a legitimation based on the internal workings of the market (as articulated in neoclassical economics) is replaced by a political legitimation with the emergence of the Keynesian welfare state and the central planning of the economy. From this point onward, political practice is measured in terms of the technical problems at hand rather than in substantive terms. The role of politics is reduced to finding the technical means to achieve goals (especially economic growth) that are understood to lie outside the realm of politics (Habermas 1970, 100–103). Technology discourse is "ideological" to the extent that political issues are treated as technical ones: tensions and contradictions are overcome by delimiting the scope of the political, and, as a result, the instrumental rationality of technical language colonizes the sphere of politics.

Habermas's formulation of technology discourse as legitimation discourse carries two arguments: a general argument regarding the depoliticizing ramifications of a technologistic consciousness and a historically specific argument regarding the legitimation of capitalism in its Keynesian stage, or, in our terminology here, in its Fordist phase. This book offers both a revival of the general argument and an updating of the historically specific argument. A revival is needed because the role of technology discourse in the legitimation of contemporary society is no less central and ubiquitous now than it has been in the past. An updating is called for because the historical circumstances that inform Habermas's thesis have since changed: both capitalism and the dominant technological paradigm have gone through sweeping transformations. It is to these transformations that I turn next.

TRANSFORMATIONS IN CAPITALISM

STRUCTURAL-TECHNOLOGICAL TRANSFORMATIONS

The last few decades have been marked by transformations in capitalism pertaining to how production is carried out, where, and by whom. According to one dominant view—offered by Régulation School theory and neo-Marxian analysis—the restructuring of capitalism has been the result of a crisis in the mode of capital accumulation that reigned in Western capitalist societies since the 1930s and came to be known as Fordism and its substitution by an emerging, more flexible mode of capitalist accumulation, post-Fordism (Aglietta 2001; Boyer 2002). Post-Fordism—the emergence of which is usually dated in the early 1970s—entails shifts from mass production to "just-in-time" production and mass customization (Harvey 1989) and from the large-scale, hierarchized, and centralized corporation that controls all aspects of production to lean, decentralized production with a core company that outsources most facets of production to other small and midsize companies (Castells 1996). The process of production has also become global, involving a global division of labor run by multinational corporations (Sklair 2002). Flexibility has come to dominate the labor process both at the micro level, with workers asked to shift between multiple tasks and ad hoc projects, and at the macro level, involving more flexible and precarious employment schemes (Greenbaum 1995; Castells 1996; Sennet 2000, 2006; Bauman 2000; Jessop 1994). Concurrent with the new regime of accumulation arose a form of social regulation that encompasses not only the mode of production but also "a facilitating shell" (Fraser 2003) of economic, social, and political arrangements, cultural and artistic sensibilities, the world of ideas and bodies of knowledge, everyday life experiences, and the conception of the individual in society. Following Gramsci (1971), we can therefore speak of a post-Fordist *society*.

The shift to a post-Fordist society is not only economic but also structural and entails a transformation in the relationships among three key actors: capital, the state, and labor. Fordist society was characterized by a strong social compact between capital and labor within the framework of a strong state. Capital made concessions to labor in the form of high wages, job security, and high levels of employment; labor conceded to capital's needs by curbing its most militant demands for a radical social revolution and providing capital with labor power for production and consumption power to purchase increasing volumes of products. The state gained political legitimacy from both groups by transforming itself into an interventionist, Keynesian, welfare state that protected capitalism from both internal failures and an external overthrow and, at the same time, provided a buffer zone between labor and the harsh realities of markets (by providing social services and installing various security schemes), hence further embedding markets in society. Appropriately, Fordist society was dominated in advanced capitalist societies by a political culture of social democracy (Harvey 2005; Polanyi 2001; Offe 1984a; Ram 2007).

Post-Fordist society is characterized by a new constellation of power among these three actors, with capital gaining increased independence from both labor and the state. This has brought about all but the demise of the Fordist social compact. Capital has become more mobile, global, and detached from particular localities and polities, leaving labor weaker. The state, likewise, lost much of its leverage to exert control over capital, which now operates on a global free market, and thus it reacted by enacting measures of self-discipline—deregulation, privatization, and downsizing. As labor became weaker *vis-à-vis* capital, it all but evaporated discursively, institutionally, and sociologically in Western societies. The power of the working class as a frame of reference for political mobilization or for the construction of political identity has been diluted as was the power of the labor movement and big unions. Collective bargaining was substituted by individual contracts, and tenured, long-term employment schemes were substituted by untenured, temporary ones (Piven and Cloward 1997; Bauman 2001; Beck 2000). Post-Fordist society is appropriately dominated by a political culture of neoliberalism (Harvey 2005).

The emergence of the new capitalism has been closely intertwined with a transformation in the dominant technology system from mechanical and centralized to informational and networked. The networked organization (Castells 1996); the command and control of headquarters in a global chain of production (Robins and Webster 1999; Harvey 1989); "just-in-time" production (Harvey 1989); the expansion in space of production, distribution, and consumption processes (Beniger 1986); and the increased flexibility of the work process (Greenbaum 1995; Robins and Webster 1999)—network technology has been instrumental in bringing about and materializing all these changes, which is precisely why so many have called ours the information, or network, society (Bell 1999; Castells 1996, 1997, 1998; Duff 2000; Lash 2002; Mackay 2003; Mattelart 2003; May 2002; Robins and Webster 1999; Webster 2002; Stehr 2001; Barney 2004).

IDEOLOGICAL TRANSFORMATIONS

The shift of capitalism to its post-Fordist phase was accompanied by the rise of a new technology discourse—indeed a new ethos and a new spirit—that wove the threads of a new technological form—network technology—into the fabric of a new capitalism and a new social regime. The transformations that are part of this new regime of accumulation of capitalism pertain not only to structural changes in the economy and in work life but also to a reconfiguration of social, cultural, artistic, scientific, emotional, political, and everyday life (Jameson 1991; Lyotard 1984). Among these is a change in the dominant political culture of advanced capitalist societies: from a political culture of social democracy to that of neoliberalism (Harvey 2005; Peters 2001; Bourdieu 1998a, 1998b; Duggan 2003; Fraser 2003). In this new constellation, the discourse on network technology plays more than a descriptive role. By filtering social reality through the prism of technology, it serves to naturalize and legitimate this reality and, in that sense, constitutes not a neutral description of reality but an ingredient thereof.

As mentioned previously, the history of technology discourse as a legitimation discourse begins with the harnessing of science and technology for capitalist development in the context of the modern state, which began roughly around the rise of industrial capitalism. The digital discourse should therefore be understood precisely in that context not only by locating it on the horizon of a *technological* change—from a mechanical and centralized industrial technology to a digital and networked postindustrial technology—and not only by locating it on the horizon of the transformation of *capitalism*—from a Fordist phase to post-Fordist phase—but also by locating it in the context of the *ideological* transformations that have accompanied these changes.

The rise of the new capitalism entails not only structural and technological transformations but also transformations in the legitimation function of technology discourse. Under conditions of a flexible regime of accumulation, globalization, neoliberalization, deregulation, privatization, and the dismantling of the welfare state, technology discourse no longer legitimates the interventionist welfare state, the central planning in businesses and the economy, the hierarchized corporation, and the tenured worker; instead it legitimates the withdrawal of the state from markets, the globalization of the economy, the dehierarchization and decentralization of businesses, and the flexibilization of production and the labor process.

This new legitimation discourse of technology marks a transformation in the "spirit of capitalism" from its industrial phase, in which the legitimation discourse emphasizes the capacity of capitalism to bring about *social emancipation* (by alleviating exploitation) to its postindustrial "spirit of networks" phase, in which the legitimation discourse focuses on capitalism's capacity to enhance *individual emancipation* (by alleviating alienation and inauthenticity and allowing more creativity and personal expression). My distinction between these two types of emancipation—social emancipation, concerned with alleviating social exploitation, and individual emancipation, concerned with alleviating personal alienation—is based on a longstanding tradition in the social sciences. It was most recently evoked by Boltanski and Chiapello (2005), who distinguish between two threads of discontent with and critique of capitalism: a social critique, focused on the harmful consequences of capitalism to the social body, that is, exploitation and inequality; and an artistic critique (variably referred to in this book as a humanist critique), focused on the harmful consequences of capitalism to the person, that is, alienation and inauthenticity. Indeed, the notion of a dual critique of capitalism is deeply rooted in the history of social thought. Marx's concept of exploitation (the extraction of surplus value) closely concurs with the artistic critique, while his notion of alienation (particularly that which emerges between workers and their labor process and between workers and their species-being) falls under the rubric of the social critique. Marcuse's notion of the deerotization of the work process and, more generally, the introduction of psychoanalysis into the analysis of modern society by the Frankfurt School informs much of the artistic critique. Weber's lament of the disenchantment of the world brought about by the iron cage, or rational instrumental bureaucracy, echoes artistic themes. Finally, in response to postmodern concerns with identity, these two types of critiques have been rearticulated as demands for redistribution and recognition (Fraser and Honneth 2003). Table 1.1 summarizes the central distinctions between these two types of critiques.

Table 1.1   Two types of critiques of capitalism

Social critique of capitalism	Artistic/humanist critique of capitalism
Exploitation	Alienation
Insecurity	Uncreativity
Lack of redistribution	Lack of recognition
Inequality	Inauthenticity
Misery	Disenchantment
Opportunism, egoism	Oppression

CAPITALISM AND LEGITIMATION

These critiques have recently been reconceptualized as central to the transformation of capitalism. In a major recent work, Luc Boltanski and Eve Chiapello register and analyze the emergence of *The New Spirit of Capitalism* (2005), thus building on and updating the classic work of Max Weber. While accepting Weber's basic framework regarding the "spirit" of capitalism, they introduce two important modifications, rendering it *historical* and *critical*. The historical dimension they introduce refers to the transformations of capitalism. Concurrent with the transformations in the operation of capitalism, its spirit changes as well. Following the framework of Régulation School theory, they detect a radical break in the spirit of capitalism between Fordism and post-Fordism. The Fordist spirit was characterized by a long-term commitment between employers and employees and by the sense of building a career path within the business organization. Businesses and corporations were clearly demarcated and distinguished from one another and from the generalized market; organizations were hierarchized and centralized; workers were expected to become highly skilled in specific tasks at which they would become increasingly proficient.

In contrast, the new spirit of capitalism that emerges with post-Fordism is based on notions of flexibility, networks, and connectionism. It entails short-term and temporal relationships between employers and employees. The locus of production becomes the network rather than autonomous companies; the axis of production becomes the ad hoc project, and work involves not specialization but the ability to multitask (see also Bauman 2001; Sennet 2000, 2006).

In addition to historicizing Weber's notion of "spirit," Boltanski and Chiapello also offer a more radicalized reading of Weber, giving it a critical edge. With his study on capitalism, Weber has taught us that capitalism entails not only the material necessities of a particular mode of production but also the idyllic features of a spirit that makes capitalism reasonable from the points of view of individual actors (Weber 1958). Between the social structure of capitalism and the agency of actors lies the realm of meaning. The spirit of capitalism endows meaning (at first, performing one's "Divine Calling" to the best of one's ability in order to gain prescience about salvation and, later, rationality) to action (the endless accumulation of capital). The spirit of capitalism provides the motivation for actors to participate in the practices of capitalism and to pursue the roles assigned to them with seriousness and intent. It provides a framework within which the mundane actions involved in the reproduction of society are given (a transcendental) meaning. In turn, it also gives capitalism legitimacy and stability. The spirit of capitalism is a discourse (to use a later term) that both describes and enables capitalism.

In its original formulation, Weber's framework explains a social change but not a critical (i.e., reasoned) change. The spirit of capitalism, according to Weber, facilitated a social change, but this change emanated from one sphere (religion) and shaped—by way of unintended consequences—the realities of another sphere (economics). Here is the second element in Boltanski and Chiapello's rendition of Weber's framework: because the spirit of capitalism is historical, it also carries with it a critical potential. By understanding the structural transformations of capitalism as juxtaposed against the substantial role of the world of ideas, Boltanski and Chiapello conceptualize the transformations of capitalism as a *response* to the critique of capitalism.

The critique of capitalism, they suggest, is inherent to the nature of capitalism itself: "a need for the unlimited accumulation of capital by *formally peaceful means*" (Boltanski and Chiapello 2005, 4; emphasis mine) also requires capitalism to make itself politically legitimate. Boltanski and Chiapello see the spirit of capitalism as a form of "justification" (Boltanski and Thevenot 2006), which explains why a particular social constellation is universally beneficial. Justification means not simply *a posteriori* legitimation of the realities of capitalism (in the way that the notion of ideology implies) but also *a priori* limitations on its *modus operandi*. This need to provide justification for capitalism creates two conditions of possibility for critical engagement and social change: capitalism might fail in the tests of its justification (i.e., it would not live up to its own promises), and, in turn, a pressure will be created to transform it either through a demand for stricter adherence to its tests or through its wholesale transformation—that is, the creation of a new justification and, in effect, a new form of capitalism.

In other words, the spirit of capitalism can also serve as a springboard for both the critique of capitalism and for its transformation. In the hands of Boltanski and Chiapello, the spirit of capitalism becomes a mediating concept between the legitimation of the prevailing constellation of power and its curbing. It lies midway between performing a discursive servitude to capitalism and offering a springboard for its critique; between acting as an ideology that disguises particularistic interests and a justification that can potentially constrain their materialization. According to Boltanski and Chiapello, the new capitalism and its spirit were constructed within the coordinates and terms set by the artistic critique of capitalism. More specifically, they claim that the new spirit of capitalism has been constructed as a response to the artistic critique of capitalism (that which laments the alienating consequences of capitalism). The upshot of this has been a suppression of the social critique of capitalism (that which laments exploitation).

WHY STUDY THE IDEOLOGY OF TECHNO-CAPITALISM?

Notwithstanding their majestic analysis of its new spirit, however, Boltanski and Chiapello completely ignore the role that technology discourse plays in the legitimation of the new capitalism. Put differently, while Boltanski and Chiapello promise to examine the "ideological changes that have accompanied recent transformations in capitalism" (2005, 3), they neglect to account for the changes in the ideology of technology that have accompanied these very transformations. This book wishes to bring technology—as a central material and ideological component—into the analysis of contemporary capitalism. It takes their framework as its entry point and registers the new spirit of capitalism as it is filtered out through the discourse on network technology, or what Castells calls "the spirit of informationalism" (Castells 1996, 195–200) and what I term the "spirit of networks."

Previous analyses have already pointed to the ideological underpinnings of contemporary technology discourse (Frank 2000; Barbrook and Cameron 1996; Best and Kellner 2000; Gere 2002; Borsook 2000; Turner 2006; Aune 2001; Mosco 2004; Wajcman 2004; Dean 2002). Such analyses tend to present contemporary technology discourse as false, utopian, or mythical. Thus, for example, Frank (2000) argues that information technology came to be "the most powerful symbolic weapon in the arsenal of market populism" (Frank 2000, 57), with the Internet specifically providing "a sort of cosmic affirmation of the principles of market populism" (Frank 2000, 79). Aune concludes that the digital discourse played a decisive role in the ideological struggle of *Selling the Free Market* (Aune

2001) during the 1980s and 1990s. Such analyses assert contemporary technology discourse to be an ideology of the free market, according to which network technology is leading to "the *apotheosis of the market*—an electronic exchange within which everybody can become a free trader" (Barbrook and Cameron 1996, emphasis mine), that is, an embodied ideal of the free market (Robins and Webster 1999, 67). Likewise, in *The Digital Sublime* (2004), Mosco understands the digital discourse as *the* myth of our time. By insisting on digital technology as ushering a historical break, it "mask[s] the continuities that make the power we observe today . . . in the global market . . . very much a deepening and extension of old forms of power" (Mosco 2004, 83).

Such works analyze the digital discourse as an ideology in the Marxist sense of the term: an ideational veil that conceals material contradictions. While such approach is insightful in its own right, it also misses an aspect of the digital discourse that I like to highlight in my analysis. Rather than assessing the *truth value* of the claims made by the digital discourse, I endeavor to decipher the particular historical form that such discourse assumes and assess its *legitimation value*. Analyzing the digital discourse as a sociological unit of analysis in its own right (i.e., as a social reality in itself rather than simply a report on social reality) will give us insight into the particular form that technology discourse assumes. Such analysis will show the dual character of contemporary technology discourse, which promises both an overcoming of the alienating components of capitalism because of its integration with network technology *and* a naturalization and depoliticization of the continuation, and even exacerbation, of its exploitative components.

By investigating the discourse on technology this book hopes to achieve two goals: first, to register the new keywords and terms, frameworks and truisms, metaphors and models that are used to account for the new lived realities of contemporary digital capitalism, and second, to explain the hegemonic, ideological, and legitimatory dimensions of this new discursive universe. In short, this book will explain the texts and subtexts of the new techno-political order of capitalist societies. The multiple social, political, and ideological coordinates of technology make its study dynamic, contingent, and historical and require the analysis of technology's specific constellations in a particular point of time. Hence, a study of the digital discourse highlights both the continuity and changes in the history of the discourse on technology in modernity.

With these particular theoretical and analytical lenses I hope to show the digital discourse to be not a simple description of the realities of network capitalism but rather a discourse that legitimates—through a "technologistic" framework (Webster and Robins 1999)—the new constellations of power entailed by the new stage of capitalism. At the center of this discourse, I argue, is the promise—anchored in network technology—of capitalism to enhance individual emancipation by alleviating alienation and the concurrent suppression of the promise to enhance social emancipation by alleviating exploitation, which has been a staple of the legitimation discourse on technology during Fordism.

CHAPTER 2

CONTEMPORARY TECHNOLOGY DISCOURSE

AFTER LOCATING IN THE PREVIOUS CHAPTER THIS RESEARCH ON contemporary technology discourse in its theoretical and sociological field, this chapter will attend to its empirical and methodological facets. The immateriality and vagueness of the notion of *the digital discourse*—the object of study in this book—demands that we first describe what it is, where it can be found, and what it looks like. The chapter then goes on to present *Wired* magazine as a suitable case study for the analysis of the digital discourse. Finally, the chapter discusses the methodological approach employed to analyze this case study.

THE DIGITAL COMMON SENSE

On May 8, 2006, the BBC News channel in Britain featured a short segment on its morning newscast concerning a court verdict in the legal struggle between Apple computers and the Apple Records label. The anchor invited an expert on network technology and the digital society, Guy Kewney, the editor of the technology Web site *Newswireless*, to discuss what this verdict "means for the industry and the growth of music online" (BBC 2006a). Let me quote this short interview with only a few omissions. It is noteworthy that the expert's mother tongue was evidently not English. The anchor opened by asking the expert, "Were you surprised by this verdict today?"

> *Expert*: I am very surprised to see this verdict to come on me, because I was not expecting that . . . So a big surprise anyway.
> *Anchor*: A big surprise, yeah, yes.
> *Expert*: Exactly.
> *Anchor*: With regards to the cost that's involved, do you think now more people will be downloading online?

Expert: Actually, if you can go everywhere you're going to see a lot of people downloading to the Internet and the Web site everything they want. But I think . . . eh . . . it is much better for development and . . . eh . . . to inform people what they want and to get the easy way and so faster if they are looking for.

Anchor: It does really seem to be the way the music industry's progressing now, that people want to go onto the website and download music.

Expert: Exactly. You can go everywhere on the cyber cafe, and you can check . . . you can go easy. It is going to be an easy way for everyone to get something to the Internet.

Anchor: Guy Kewney, thanks very much indeed. (BBC 2006a)

That these few blurbs by an expert on your average morning newscast do not sound particularity outrageous, surprising, or radical is precisely the punch line of this little anecdote: due to a "casting" mishap, the interviewee was not in fact Guy Kewney the expert, but Guy Goma, an immigrant from Brazzaville, Republic of the Congo, who arrived to the BBC building to be interviewed for the position of "Data Support Cleanser" in the IT department and was called in to the morning show studio by mistake. In an apology broadcasted later by the BBC, the anchor expressed what was probably a common sentiment among viewers, saying that "to his [Guy Goma's] credit, as the interview continued he had a good stab at giving some answers" (BBC 2006b). That Guy Goma could rather easily pass as an expert on contemporary digital culture is a testament (anecdotal nonetheless) to how ubiquitous and hegemonic the discourse on network technology has become. His answers sounded reasonable because they complied with the hegemonic discourse on network technology (would any of us give different answers?). Guy Goma said what we already know; he had spoken the discourse on contemporary network technology, or the digital discourse.

THE DIGITAL DISCOURSE

The digital discourse is a body of knowledge aimed at making sense of contemporary society. The premise of this discourse, prevalent in academic, business, political, and popular circles, is that the development and widespread deployment of network technology has ushered in a new society and a new civilization that is so revolutionary that established discourses fail to comprehend and represent it; hence the need for a new discourse. The unique characteristics of network technology are seen as paradigmatically different from previous technological systems, and hence they are seen as socially transformative. The general tone of the digital discourse toward these transformations is usually located on the spectrum between optimistic and euphoric. While problems are acknowledged, the overall movement is seen as universally progressive (with problems promised to be resolved, usually by more sophisticated technology). Thus, a digital civilization is ultimately a more benevolent civilization. The metanarrative of the digital discourse is one of revival: of the economy, of political life, of sociability, and of culture.

The scope and prevalence of the discourse on network technology as a major protagonist of social life is evident even in a cursory skimming through contemporary culture. Indeed, that such a superficial survey brings such a wealth of information from so many cultural spheres is perhaps the best testament to the prevalence of the discourse: it is mundane, quotidian, a matter of fact, and common sense. Let me point out just a few of the manifestations of the digital discourse in the public sphere.

The staples of digital culture—its heroes, artifacts, and institutions—have become part of the general popular culture. Bill Gates, iPod, Google, and YouTube exert influence far beyond their professional boundaries or functional value; they have become iconic of the digital era. They are no longer part of a "geek" culture or the IT professional elite but are at the heart of contemporary culture. Likewise, much of the terminology surrounding network technology—"high-tech," "dot com," "bugs," "open source," "networking," and many more—has percolated into the general public discourse. A striking example of that is the distinction between "hardware" and "software," which is drawn straight out of the technical terminology of computer professionals and has become a metaphor and a powerful cognitive model in a variety of spheres and discourses.

Guy Goma aside, experts specializing in the new social realm of network technology, cyberculture, and so forth have indeed flourished. These experts and self-defined gurus were sprouting on the public sphere, illuminating for the public the new technological realities and their social ramifications. A reliable indication of the popularity of the digital discourse is the proliferation of nonfiction books aimed at the general public. A list of some of the more popular titles published recently illustrates the extent to which network technology has come to dominate public discussion and has become a public matter. The list also underscores the general tone of optimism prevalent in the discourse: technology is mostly seen as a tool for emancipation and liberation, for overcoming obstacles and amending problems, and for change for the better.

The narratives of the digital discourse, which will be unfolded at length in the following chapters, have been explored in numerous popular books:

- The narrative of a technologically induced revolution dominates titles such as *The Evolution of Wired Life: From the Alphabet to the Soul-Catcher Chip—How Information Technologies Change Our World* (Jonscher 1999).
- The idea of human emancipation and even transcendence through network technology has been explored in such books as *The Singularity Is Near: When Humans Transcend Biology* (Kurzweil 2005), *The Age of Spiritual Machines: When Computers Exceed Human Intelligence* (Kurzweil 2000), and *The End of Medicine: How Silicon Valley (and Naked Mice) Will Reboot Your Doctor* (Kessler 2006).
- The narrative of individual empowerment *vis-à-vis* the state and the market has been discussed in *An Army of Davids: How Markets and Technology Empower Ordinary People to Beat Big Media, Big Government, and Other Goliaths* (Reynolds 2006), *The Power Of Many: How The Living Web Is Transforming Politics, Business, And Everyday Life* (Crumlish 2004), and *Me++: The Cyborg Self and the Networked City* (Mitchell 2004).
- The narrative of economic revitalization is captured in such titles as *The Weightless World: Thriving in the Digital Age* (Coyle 1999), *The Long Tail: Why the Future of Business Is Selling Less of More* (Anderson 2006a), *The Wealth of Networks: How Social Production Transforms Markets and Freedom* (Benkler 2006), *Naked Conversations: How Blogs are Changing the Way Businesses Talk with Customers* (Scoble and Israel 2006) and *The Search: How Google and Its Rivals Rewrote the Rules of Business and Transformed Our Culture* (Battelle 2005).
- Networks, their tremendous positive potential, and the viral spread of networks from technology to society has been explored in *Emergence: The Connected Lives of Ants, Brains, Cities, and Software* (Johnson 2002), *Nexus: Small Worlds and the Groundbreaking Science of Networks* (Buchanan 2002), *Linked: How Everything Is Connected to Everything Else and What It Means* (Barabasi 2003), *Smart Mobs: The Next Social Revolution* (Rheingold 2002), *The Moment of Complexity: Emerging Network Culture* (Taylor 2003) and *Peer-to-Peer: Harnessing the Power of Disruptive Technologies* (Oram 2001).
- The chaotic character of networks has been famously explored in *Out of Control: The New Biology of Machines, Social Systems and the Economic World* (Kelly 1995).
- Blogs and their positive effects on individual empowerment and the revitalization of civic life are explored in such titles as *We've Got*

Blog: How Weblogs Are Changing Our Culture (The Editors of Perseus Publishing 2002), *Who Let the Blogs Out?: A Hyperconnected Peek at the World of Weblogs* (Stone 2004), and *Blog!: How the Newest Media Revolution is Changing Politics, Business, and Culture* (Kline and Burstein 2005).

Another reliable indicator of the ubiquity of the digital discourse and its penetration into general public discussion can be glanced from the covers of *Time* magazine. A survey of covers of the magazine between 1980 and 2008 reveals more than forty covers featuring network technology and its human and nonhuman protagonists. (This list includes covers that refer specifically to digital technology and excludes covers that refer to technology and science in general, such as medical technology or space travel.) The significance of this survey is that network technology appears as the cover story in one of America's most popular and influential weekly newsmagazines. In these issues, digital technology and its protagonists trump politicians and various social issues (such as school violence or sex) as the top general interest concern of the day. Three covers of special importance are those featuring *Time*'s "Man [and later, "Person"] of the Year." The 1982 Man of the Year was the personal computer (PC). Andy Grove of Intel was man of the year for 1997. And the 2006 Person of the Year was "You," that is, each and every individual user of network technology.

WIRED MAGAZINE AS AN EPITOME OF THE DIGITAL DISCOURSE

To study the hypothetical construct of the digital discourse, I will delimit myself mostly to the analysis of a single, well-demarcated, and concrete articulation of the digital discourse that would serve as a case study: *Wired* magazine, one of the most consistent and crystallized articulations of the digital discourse. Published since 1993 with great success, *Wired* has issued more than 180 monthly issues thus far and has a current circulation of approximately seven hundred thousand (Darrell 2007). *Wired* grew out of, and indeed concurrently with, the dot-com bubble, the explosion of network technology and the Internet, and the emergence of the high-tech industry as a central engine in the economy and a powerful source of identity and culture. It is located, socially and geographically, at the heart of the revolution on which it purports to report: new technologies, new businesses, new people, and new media that make up the digital universe. *Wired*, in other words, is also part of that revolution. As the magazine's founder puts it, *Wired* is "the mouthpiece of the digital revolution" (Wolf 2003, 52).

In this research, *Wired* is seen as a key articulator of a discourse that not only reflects the realities of a new digital capitalism but also endows them with a "spirit" and makes sense of new constellations of power and new forms of life. The digital discourse is a "historically, socially, and institutionally specific structure of statements, terms, categories, and beliefs," perceived as offering a transhistorical "truth"; "Precisely because they are assigned the status of objective knowledge, they seem to be beyond dispute and thus serve a powerful legitimating function" (Scott 1988). The purpose of this work is to uncover the historical, social, and institutional *specificity* of these new statements, terms, categories, images, and beliefs about contemporary capitalist society as they are crystallized in the digital discourse and point to their legitimation function.

Utilizing *Wired* magazine as a case study will allow me to expose and penetrate the outer shell of the black box of the digital discourse. A case study is a research model that focuses on a singular social entity or cultural artifact, treats it as a bounded system, and utilizes various methods to study it (Marshall 1994; Stake 1994). A case study is a naturally occurring social phenomenon, selected not chiefly for its typicality, but on the basis of its theoretical significance. The purpose of a case study is to provide the materials for what Geertz (1977) calls a "thick description," which is essential to reach an understanding of the broader context (Kuper and Kuper 1996). On the other side of the coin of a thick, deep, and meaningful description is the question of how well the case study represents the sociological phenomenon at hand. This makes the selection of the case to be studied critical. A case study must find the global within the local, the general within the specific (Hamel, Dufour, and Fortin 1991, v, 36–38).

Wired is precisely this unique representative of a broader, social digital discourse. It is unique in the sense that it represents this discourse at its most concentrated and crystallized form. It is a more valuable case study than other publications that are part of this public discourse in that it is committed wholly and solely to this particular subject. The value of a case study is measured by how much it facilitates analytical, rather than statistical, generalization (Yin 1989, 21). So while *Wired* can correctly be characterized as a more extreme case than, say, the *New York Times* in both quantitative and qualitative terms (i.e., in how much it deals with network technology and in what ways), it is precisely these characteristics of being explicit, straightforward, and mobilized to the digital "cause" that yields more fruitful and cogent results.

This, however, is not a book about *Wired*; the magazine serves merely as an "instrumental case study" (Stake 1994). The case itself—notwithstanding the deep analysis it receives—is only of secondary significance;

the goal is ultimately to be able to answer the theoretical question regarding the legitimation role of technology in the political culture of advanced capitalist societies. The methodology of deciphering a dominant discourse through the reading of popular magazines has been extensively utilized in the social sciences. For example, *Reading National Geographic* (Lutz and Collins 1993) studied national, racial, gender, and geographical representations in the United States through a close textual analysis of the photography featured in the famous magazine. Edward Said famously studied how news outlets in the United States are *Covering Islam* (1981). And more recently, a group of scholars has offered an analysis of masculinity, capitalism, and consumption in contemporary Western society through reading and *Making Sense of Men's Magazines* (Stevenson, Jackson, and Brooks 2001). In this book, too, the analysis of the magazine takes on a supportive role to facilitate understanding of another, broader issue. Also, since the close observation of the case (*Wired* magazine) helps us understand an external phenomenon (the digital discourse), the case study does not have to be typical of other cases as long as it increases our understanding of the wider phenomenon (Stake 1994).

Wired was founded by American journalists Louis Rossetto and Jane Metcalfe. It was a great success at its launch and was compared to *Rolling Stone* magazine for its innovation and cultural impact. It won five National Magazine Awards for General Excellence and three for Design (American Society of Magazine Editors n.d.). The magazine brought together a varied cadre of individuals, such as leaders from the *Electronic Frontier Foundation* (a techno-libertarian advocacy group), academics from the Stanford Research Institute, Nicholas Negroponte from the Massachusetts Institute of Technology Media Lab, and Kevin Kelly, editor of *Whole Earth Review* and author of *Out of Control: The Rise of NeoBoiological Civilization* (1995; Wolf 2003, 48, 50, 53). The company included not only the monthly magazine but also dozens of Web sites, a search engine (Hotbot), a blog, and a stock index. After a failed attempt to take the company public in 1997, it was sold off in pieces. *Wired* magazine was purchased by Advance Magazine Publishers, who assigned it to its New York subsidiary, Condé Nast. The editorial offices were kept in San Francisco, close to the epicenter of the digital revolution—Silicon Valley.

Wired perceived what was happening at that epicenter as the bedrock of a social revolution, a revolution that is ignored by mainstream media, which continued to concentrate on the old, industrial-era structures of power and social dynamics. The magazine's goal was to register and articulate this new revolution and make sense of the new society to which it gives birth. *Wired*'s first issue makes explicit some of the motivations of the magazine's introduction and some of the themes that will dominate it for years to come. The alignment around the axis of new media and technology was laid out in the magazine's first words, taken from Marshall McLuhan's 1967 book *The Medium Is the Message*:[1] "The medium, or process, of our time—electric technology—is reshaping and restructuring patterns of social interdependence and every aspect of our personal life. It is forcing us to reconsider and re-evaluate practically every thought, every action, and every institution formerly taken for granted. Everything is changing: your education, your family, your neighborhood, your job, your relation to 'the others.' And they're changing dramatically" (McLuhan 1967, cited in Wolf 2003, 71). In the next page of its inauguration issue, cofounder Louis Rossetto asserts why *Wired* is needed: "Because the Digital Revolution is whipping through our lives like a Bengali Typhoon—while the mainstream media is still groping for the snooze button. And because the computer 'press' is too busy churning out the latest PCInfoComputingCorporateWorld iteration for its ad sales formula cum parts catalog to discuss the meaning or context of social changes so profound their only parallel is probably the discovery of fire" (*Wired*, January 1993, cited in Wolf 2003, 71).

In the January 1998 issue, celebrating the magazine's five-year anniversary, Rossetto reiterates the magazine's mission. He recalls his difficulty to pitch to investors the notion of a revolution brought about by information technology: "Digital Revolution? Most publishers were barely aware of the Industrial Revolution . . . What we were dreaming about was profound global transformation. We wanted to tell the story of the companies, the ideas, and especially the people making the Digital Revolution. Our heroes weren't politicians and generals or priests and pundits, but those creating and using technology and networks in their professional and private lives—you" (Rossetto 1998).

From the heights of the dot-com euphoria, Rossetto outlines the magazine's basic theory of contemporary society and the role of the magazine in it:

> The Internet has mushroomed from an obscure academic mail system into the fastest growing medium, marketplace, and community in history. Genetic engineering is conquering disease, and new energy technologies promise to save our environment. The global financial network has created a force for change more powerful than the nation-state. And digital citizens are reinvigorating democratic discourse and reinventing civil society . . . We at *Wired* remain obsessed with authoritatively reporting on the new economy, new media, crucial technologies, and the digital nation. With providing context for a community overwhelmed by data. With not just telling the story of the modern, but visualizing the excitement of our times. (Rossetto 1998)

Right from the outset *Wired's* founding rationale is that the technological revolution is yielding a social revolution. Hence, the main protagonist of the magazine is network technology itself, as well as the people and organizations intimately engaged with it. But *Wired* is not a magazine about technology but about society, only every topic is filtered through network technology. The stories are told through technology, but they are really about every conceivable facet of society. Later, other, more traditional publications, such as the *New York Times* and *Newsweek*, followed suit with their own technology sections. This shift from news stories that involve network technology to a weekly section devoted to (mostly digital) technology signifies the extent to which network technology entered as a subject—right next to the economy and politics—of everyday, generalized public discourse. "Network technology" has become a staple of our political discourse.

In fact, as Gary Wolf—an intimate witness to the formation of the magazine and a contributing editor—suggests, the roots of this digital discourse in the public sphere had sprouted before the inauguration of *Wired*; there was just a need for someone to crystallize the discourse and to consolidate its fragmentary voices into one harmonious sound. In that respect, the magazine sprouted unto a fertile ground. Recalling work on the pilot issue of the magazine, Wolf writes in *Wired: A Romance* (a biography of the magazine and its creators), "None of the articles or photographs were [*sic*] original . . . They were cribbed from every type of publication that touched on parts of the story [cofounders] Louis and Jane were trying to tell," types of publications such as the *Wall Street Journal*, the *New York Times*, *Business Week*, and numerous others. "But," Wolf continues, "the borrowed stories seemed bolder here than they have in their original publication. Bound together, they radiated a sense of fanatical self-assurance, as if united in expectation of technological wonders and tremendous social change. The fact that the stories were taken from mainstream sources strengthened rather than softened the effect, for the credits page offered evidence that the revolution Louis believed he was chronicling has already been widely noted only the context was new" (Wolf 2003, 47). *Wired* then was serving as a reverse prism: unifying the variety of representations of an emerging digital discourse in the public sphere into a coherent, cogent, and bright voice.

My argument is that *Wired* is representative of a much broader social phenomenon. At the same time, it has several unique characteristics that distinguish it from three other types of publications that also focus on technology and that makes it particularly suitable for studying contemporary technology discourse. The first type is trade magazines, such as *PC World*, *PC Magazine*, and *InfoWorld*, which are geared toward professionals in the field. These publications center more on the technical facet of information technology. Indeed there were some other professional publications (such as *Mondo2000* and *Signal*) that also grew out of the epicenter of the high-tech industry (in and around San Francisco) that tried to capture the same technological and social revolution that *Wired* was after. However, while other technologically centered publications targeted specialized and professional audiences, *Wired*'s goal was "to broaden the definition of the digital vanguard to include hundreds of thousands of readers" (Wolf 2003, 49). In that respect, *Wired* thought the digital revolution to be a deep and populist social revolution brought about by individual users rather than by experts or hackers. Thus, *Wired* saw its goal in providing these individuals monthly updates on this emerging world—*their* world (Wolf 2003, 50).

The second type of publication from which *Wired* needs to be distinguished is comprised of publications like *Technology Review*, *Spectrum*, *Discover*, *Popular Science*, and *Scientific American*, which are aimed at the general public. They usually present a discourse of fascination of, and even fetishization of, network technology, but, unlike *Wired*, they see their value mostly in bringing technological news to society and rarely venture into social, political, and cultural questions. Another advantage of *Wired* compared to this second group is the former's concentration almost exclusively on network technology, rather than technology (and sometimes science) in general, and its integration within the economy and culture of Silicon Valley. To sum up, unlike those two types of publications, I see *Wired* not as a magazine about technology but rather as a magazine about the intersection of technology and society. It is oriented not toward a professional public and technical issues, but toward the general public and social issues. It is a current affairs magazine, just like *The Economist*, *Time*, and *Newsweek*, only it perceives *technology* (rather than politics or economics) to be the predominant axis of these affairs in contemporary society.

That leaves us with the task of differentiating *Wired* from weekly publications on current affairs (such as *Time* or *The Economist*) as well as daily newspapers, all of which now regularly feature a special section on technology. Here the distinction is much more subtle, and the choice of *Wired* rather than the other publications is mostly methodological. First, in terms of content, *Wired*'s discourse tends to be what we might call more "ideological." The digital discourse, which is the hypothetical structure that this book seeks to capture, can be found in a more crystallized and pure form in *Wired* rather than in, say, the *New York Times*. Second, while *Newsweek* or *Time* are established magazines that have integrated a technology section into their older content, *Wired* is *sui generis*, that is, it was created precisely for the sake of registering and defining the digital discourse. It preceded some of those other publications by a few years in terms of a technologically centered reporting, and, in terms of social networks and geography, it has been located at the heart of the revolution on which it purports to report.

DISCOURSE AND ITS ANALYSIS

The notion of discourse is a central concept in this book, both theoretically and methodologically. Chapter 1 made a theoretical claim and asserted the discourse on technology in general and the digital discourse in particular as valid and important sociological objects of study. Let me add here a more general discussion on discourse and consider in particular the interrelations between discourse and practice. Since the linguistic (or interpretive) turn, language has come to be seen as a central arena of social production. Language is a system "through which meaning is constructed and cultural practices organized and by which, accordingly, people represent and understand their world" (Scott 1988). Language is neither a cause nor a reflection of social relations, as the idealist-materialist opposition implies. Rather, "the analysis of language provides a crucial point of entry . . . for understanding how social relations are conceived" and, in turn, "how institutions are organized, how relations of production are experienced, and how collective identity is established" (Scott 1988).

Whereas structuralist analysis seeks to uncover the hidden meaning of language—that is, to reduce the multiplicity of the phenomenology of language (the superstructure) into the unitary and essential meaning (the base)—the assumption behind the poststructuralist approach is that society takes place precisely within language and discourse. Language generates meanings and practices that constitute the social. For Foucault, discourse involves the production of knowledge that "gives meaning to material objects and social practices" (Barker 2003, 102).

While the knowledge produced by discourse does not reveal any essential and ultimate truth, it is not arbitrary; Foucault insists that discourse "generates meaning under specific material and historical conditions" (Barker 2003, 101). The task of the researcher is therefore not to examine the link between discourse and reality but to point out the link between discourse and its conditions of possibility—that is, the social conditions under which it has been produced—and the link between discourse and its effects.

Chouliaraki and Fairclough (1999) assert the significance of the study of discourse as part of the project of critical social science, especially under contemporary political and economic conditions. Contemporary social changes, with their global scale and sheer complexity, "increase the sense of helplessness and incomprehension" (3). "Social forms that are produced by people and can be changed by people are being seen as if they were part of nature" (4); they are therefore "widely perceived as inevitable" (3). The analytical category of discourse can be critical insofar as it highlights that discourse and practice are mutually constitutive. Bourdieu (1998a, 1998b) points out that the notion of "flexibility," central to the neoliberal discourse, is already a reality in global capitalism, "but it is also backed by social forces (e.g., the banks) which aim to make it more of a reality, and the discourse of flexibility is one of the [symbolic] resources they have . . . for achieving this" (Chouliaraki and Fairclough 1999, 4). The relationship between discourse and social practice is dialectical so that "imaginaries," that is, "representations of how things might or could or should be" (Chiapello and Fairclough 2002) are enacted as practices, such as technologies, bodies of knowledge, genres implicated with social practices, and identities, which in turn become the material for further representations of reality. In the language of Bourdieu this book aims to decipher and deconstruct the digital discourse as a "symbolic resource," which partakes in the construction of the digital society.

CONCLUSION: DISCOURSE, IDEOLOGY, AND SPIRIT

The central category analyzed in this study is that of discourse, but there are two additional, closely related concepts that are discussed in the book: ideology and spirit. The genealogy of ideology, spirit, and discourse in the history of ideas is as diverse as the history of social thought itself. Ideology is traced back to Marx, spirit to Weber, and discourse to Foucault. Ideology is a structuralist concept, spirit focuses on agency, and discourse is derived from poststructuralist thought. Ideology points to a divergence between truth and (both common and false) knowledge; spirit—to the meanings that shape social action; discourse—to knowledge that becomes truth through practice and power. But they all share a common sensibility that is at the heart of this book. Discourse, ideology, and spirit point to the social function of knowledge; in all of them knowledge is treated not simply as a function of cognitive reflection but as closely intertwined with social practices.

To some degree, discourse, ideology, and spirit—as well as hegemony (Gramsci 1971), mythology (Lévi-Strauss 1963; Barthes 1972), and doxa (Bourdieu 2003)—engage the legitimation function of knowledge. They all point to the characteristics of a particular type of knowledge that is socially embedded, intertwined with practice, self-evident, *a priori*, and axiomatic. Such knowledge is resistant to scrutiny precisely because it is woven into deeper constructs, such as an "outlook," or "identity" (Ram

2006, 13), or as Raymond Williams puts it, into the "structure of feelings" (Williams 1978). This type of "natural" knowledge is the backbone of political culture and political legitimacy. As Lyn Hunt argues, "The legitimacy of political authority depends on its resonance with more global, even cosmic cultural presuppositions, for political life is 'enfolded' in general conceptions of how reality is put together" (Hunt 1984, 87). "The digital society," "the network society," and so forth are "master fictions" (Geertz 1977) that bring together these various presuppositions and endow them with shared meaning (Ram 2006, 13). The purpose of this book is to analyze the digital discourse as this kind of knowledge and show the usually invisible threads that connect it to politics and power.

PART II

CHAPTER 3

NETWORK MARKET

PART II OF THE BOOK PRESENTS THE ANALYSIS OF the digital discourse at the intersection of network technology and the new, post-Fordist capitalism. Over the course of four chapters, it presents the central narratives of the digital discourse and underscores how these constitute a new "spirit of networks." As the market, work, production, and the human become integrated into network technology, so are their essences transformed. The next four chapters analyze the digital discourse on the network market, network work, network production, and the network human.

As outlined in Chapter 1, beginning in the 1970s, capitalism has gone through radical transformations in its mode of operation. From a functionalist point of view, these internal transformations were necessary in order to revitalize capitalism and represented a response to its inherent contradictions and crisis tendencies (Mandel 1978), which at that time have reached their zenith and threatened to collapse capitalism onto itself. The ultimate goal of capitalism as a world system has remained the same: the infinite accumulation of capital by peaceful means (Wallerstein 2004), but its means, or its mode of accumulation, was revolutionized, shifting from Fordism to post-Fordism.

THE NEW DYNAMICS OF MARKETS IN POST-FORDISM

One of the focal points for this transformation has been the market: the emergence of post-Fordism entailed a radically new constellation of power between the market and the state and, more profoundly, between the market and society, with markets becoming increasingly disembedded from society (Polanyi 2001; Harvey 2005). This disembeddedness—part of a broader social transformation from Fordism to post-Fordism—is dominated by two trends: marketization and disorganization.

Marketization entails the increasing dominance and scope of markets in social life. Markets have gained more autonomy *vis-à-vis* the state, becoming more deregulated and more globalized (Castells 1996; Sassen

1999). The state withdrew not only from intervening in the workings of the market but also from ownership of *The Commanding Heights* of the economy through privatization (Yergin and Stanislaw 1998). The state also withdrew from the funding and operation of many welfare mechanisms that were put in place in order to provide a buffer zone between individuals and the market (Piven and Cloward 1997). More and more spheres of social life are administered by the free market or were modeled after a market-like rationale (Somers and Block 2005). There has been a trend toward privatization by way of shifting risks and responsibility from the state to individuals, and there has also been a process of privatization in the world of work, where a class compact has been substituted by individual contracts. The decline in market regulation and income redistribution (or in fact the rise of upward redistribution [Duggan 2003]) has also led to an increase in class inequality within national boundaries and between nation states (Milanovic 2007; Harvey 2005, 2009).

The second trend characteristic of the disembeddedness of markets in post-Fordism is disorganization. *Disorganization* (Lash and Urry 1987; Offe 1984b)—partially a consequence of the marketization of society—refers to a process whereby markets, the economy, and social life in general have become more liquid (Bauman 2000), more chaotic, and more complex (Urry 2003). The globalization of financial markets has made capital more mobile, leaving local markets more volatile and unstable as a result (Sassen 1999; Sennet 2006; Harvey 2005). Production has become more flexible, constantly adapting to changing markets' demands; production and consumption cycles have been accelerating (Harvey 1989; Agger 2004; Rosa 2003). Companies have shifted their organization from a vertical model of a top-down, hierarchized bureaucracy to a horizontal, dehierarchized, and decentralized network (Castells 1996; Sennet 2006). Flexible, lean, just-in-time production has made work life more "mean" (Harrison 1997), and increasingly more precarious, unstable, and unpredictable (Bauman 2001). Tenured workers have been replaced by part-timers and flextimers, working on ad hoc projects rather than developing a linear career path (Castells 1996; Sennet 2000). Economic risks (as well as spoils) have been individualized (Bauman 1998, 2001; Beck 1992, 2000).

These two dynamics were brought about not only as a result of the structural constrains of capitalism but also by the force of an ideological and political struggle between Keynesianism, welfarism, embedded liberalism, and social democracy, on the one hand, and market fundamentalism and neoliberalism, on the other (Somers and Block 2005; Duggan 2003). One of the staples of Keynesianism involved tighter state regulation of its macroeconomic indicators. This meant that while markets remained essentially free, they were further embedded in society (Polanyi 2001). This had given us a period of four decades of embedded liberalism (Harvey 2005), or "traditional mixed economy" (Yergin and Stanislaw 1998, 127). Diminishing state revenues in the 1970s mark the beginning of increasing skepticism of that constellation. A new model appeared that undermined the central tenets of Keynesianism and sought to solve the inherent contradictions of the welfare state (Offe 1984a).

Since Thatcherism in the United Kingdom was one of the original formulations of the new economic doxa of neoliberalism, it is worthwhile recounting its history here. The neoliberal reformers in Britain identified the source of the economic crisis of the 1970s as "'thirty years of socialistic fashions' and 'statism'—three decades of looking to government to solve problems and run the economy" (Yergin and Stanislaw 1998, 99). The new objective for the government, they asserted, should be the creation of wealth, even at the cost of creating inequality and mass unemployment (Yergin and Stanislaw 1998, 101–2). In Britain, this mostly took the form of privatization, a measure that not only had immediate policy consequences but also was part of a larger political and ideological project of transforming Fordist society and its political culture. Privatization—the transfer of industries and services from state ownership and management to the market—was seen as a means of "changing the balance in society" (Yergin and Stanislaw 1998, 115). It represented a critique of the notion that centralized "government knowledge" is superior to decentralized "business (or market) knowledge." Moreover, privatization was seen as contributing to an ideology of individualism aimed at "overturning the previous collectivist consensus and replacing it with free-market economics and rigorous individualism" (Frith 1994, 1). Assessing the legacy of Thatcherism, Yergin and Stanislaw conclude that "Thatcherism shifted the emphasis from state responsibility to individual responsibility, and sough to give first priority to initiative, incentives, and wealth generation rather than redistribution and equality. It celebrated entrepreneurship . . . By the end of 1990s, it would turn out that Margaret Thatcher had established the new economic agenda around the world" (Yergin and Stanislaw 1998, 123).

This new economic agenda—neoliberalism—not only has become the dominant economic policy in Western capitalist societies (and increasingly a global doxa) but also has helped shape a new political culture in these societies. As an academic and popular discourse, neoliberalism gave a legitimation for the new, post-Fordist constellation of power between capital, labor, and the state and, especially, reasserted the need to liberate the market from political oversight and integrate social life as much as possible into markets. In this transformation, the role of the digital discourse was paramount.

This chapter examines the digital discourse at the intersection of network technology and the market. It introduces key texts in *Wired* magazine that deal with contemporary economy and specifically with the transformations in the workings of the free market due to network technology. According to the digital discourse, as the market is integrated into network technology, it becomes more frictionless and more rational; it is rendered more self-regulating and hence demands almost no external political regulation; and it becomes more chaotic and unstable and, in turn, requires the nodes of the network—both individuals and companies—to be flexible and adaptive to an ever-changing environment. These narratives of a network market provide a cognitive map that explains the transformation from the embedded liberalism of the Fordist era to the neoliberalism of the post-Fordist era. By relying on a technologistic framework, the digital discourse on the network market legitimates the trend whereby markets are becoming more autonomous and disembedded from social regulation and are seen as ideal mechanisms for the administration of social reproduction. Through a close reading of the similarities between the digital discourse and neoliberal discourse, this chapter uncovers the degree to which the notion of network markets in the digital discourse is founded on the basic image of markets as portrayed by neoliberal theory. The notion of networks is rooted and constructed within the paradigm of markets—that is, as a template to understand not only markets but also society *as* markets. Set within the theoretical framework of this book, this chapter argues that the discourse on network markets legitimizes and depoliticizes the withdrawal of the Keynesian welfare state and the emergence of a new economic regime of neoliberalism. This chapter also points out the degree to which the network market is seen as a response to the humanist critique of capitalism and a transcendence of the limitations set forth by the embedded market of the Fordist era.

The chapter proceeds by analyzing two key dynamics that lie at the heart of the digital-discourse explanation of the revolutionary nature of the network economy: spontaneous order and chaos. According to the digital discourse, the new economic and market rationality of the network economy relies on spontaneous order, which in turn is premised on network technology. The contemporary realities of flux, instability, unpredictability, and chaos in the market are tied in the digital discourse to the market being more networked and its operation based on spontaneous order. The last part of the chapter adds another dimension to the analysis by discussing the theoretical underpinning of neoliberalism. Through a discussion on spontaneous order and chaos in neoliberal theory, this chapter shows the transposition of a theoretical model and a political project into a technologistic and seemingly apolitical discourse. The concluding discussion clarifies the theoretical significance of the affinity between neoliberalism as a market ideology and the discourse on network technology within Habermas's framework of a technologistic political legitimation.

SPONTANEOUS ORDER AS NETWORK RATIONALITY

DUMB NODES, SMART NETWORK

One of the foundational texts describing (as well as prescribing) the new workings of the network market is "New Rules for the New Economy" (Kelly 1997)[1] by Kevin Kelly, *Wired*'s editor in chief. I will focus my analysis of the discourse on the network market mostly on a close reading of this text, as well as some other articles from *Wired*. The most exciting feature of the digital revolution, according to Kelly, is the emergence of the network as the prominent paradigm of technological architecture and, in turn, of economic activity. As Kelly puts it, "The new economy is about *communication*" (1998, 5)[2]; specifically, it is about the ability of disparate units to be networked or of individual nodes to communicate amongst themselves.

Kelly begins his analysis of the economy with a close examination of its underlying technology; a technological analysis that is supposed to unravel the shape of contemporary society. He notes that "dumb chips"—simple processors designed to perform very limited computational tasks—are becoming much more popular than the more sophisticated computer chips (Kelly 1998, 10–11). Computer chips (such as the central processing units [CPUs] within PCs) are characterized by self-sufficiency; that is, they are stand-alone, autonomous units. Dump chips, on the other hand, only make sense within a network structure. Each of these chips is "dumb," but as we are "connecting these billion nodes, one by one" (12), these small, not-so-intelligent machines become something else; they gain new qualities of "smartness" (14) and rationality (16).

Such a network is characterized by a high level of rationality that is brought about not by any single supercomputer, which in itself is external to the network and controls its operation—an omniscient eye in the sky, or a Big Brother that watches over the network. Instead, network rationality is brought about by the mere interlinking of dumb chips—and more generally, individual nodes—into a web. Intelligence, knowledge, and economic rationality, according to Kelly, reside not in any individual node but only in the network, and it comes about only through the new technological ability of nodes to come together in a collective rational action—that is, to "swarm" (12). Order and rationality are brought about by the interlinking of simple, irrational nodes. This type of order, Kelly says, emerges in any system that employs network architecture—biological, technological, economic, cultural, and social. Thus, Kelly is able to extrapolate from the technological level to other realms, such as intelligence. For example, he asserts that "when connected into a swarm, small thoughts become smart" (12). The interconnection of many small, simple-minded parts results in a qualitative leap—so that "small" becomes not simply big but "smart." Using the same model, Kelly explains human consciousness as the communicative cooperation of dumb neurons: "Our brains," Kelly says, "tap into dumb power by clumping dumb neurons into consciousness" (13). Networks are mechanisms for radical qualitative transformations and for achieving a gestalt, or a whole that is bigger than the sum of its parts.

It is important to infer what Kelly is suggesting, especially as it pertains to the status of nodes *vis-à-vis* a technological network, or that of individuals *vis-à-vis* markets and society. If consciousness (as well as smartness and rationality) is the result of the cooperation of dumb neurons (as well as dumb chips, nodes, or individuals), the corollary is that *reflexivity* (i.e., the ability to evaluate rationality, or the ability to apply rationality to asses rationality [see Beck, Giddens, and Lash 1994]) resides not in any single node but only in the network. None of these small nodes can comprehend the complexity of the network's rationality. Kelly sums up this lesson by maintaining that "no one is as smart as everyone" (Kelly 1998, 13); he therefore redefines smartness as an inherently network quality. This simple technological premise regarding the inability of any one node to comprehend the complexity of the web would, as we will shortly see, prove to be consequential when explaining actors in markets and the futile attempt of any agency to comprehend markets, let alone control them.

A NEW KIND OF ORDER

How does network rationality come about? According to Kelly, as long as disparate units are interconnected by technologically enabled networks, rationality and order will emerge spontaneously. After citing evidence for the emergence of spontaneous intelligence and rationality in networks from a variety of case studies in technology and nature Kelly draws two general rules—rules, we should be reminded by the title of his article and book—for the new economy: "Dumb parts, properly connected into a swarm, yield smart results" (Kelly 1998, 13); and "The surest way to smartness is through massive dumbness" (14). Put together, these rules suggest that the network is the best mechanism to produce rationality. Moreover, it suggests that superior rationality can solely be the product of networks. Intelligence and rationality are achieved not by improving on the performance of individual nodes but by connecting them to each other. Sophistication and progress is created by very limited, short-sighted, and unreflexive agents. This network vision of rationality heralds, according to Kelly, the death of an alternative—and once hegemonic—vision of rationality encapsulated in the notion of artificial intelligence: a centralized, self-sufficient, and omniscient supercomputer (14). Rationality, in conclusion, involves two elements: dumb nodes and the mechanism that connects them and self-regulates their action. The Internet and other network technology serve not simply as the quintessential metaphor for this new rationality but indeed as the material basis for its execution.

The various concepts used to account for this new form of network architecture and network rationality in the digital discourse is very telling. "Distributed power" (Anderson 2002, Luman 2005, Leslie 1997), "smart mobs" (Rheingold 2003, Sterling 2005, Kelly 2005), "the wisdom of crowds" (Surowiecki 2004), "swarm intelligence" (Bonabeau, Dorigo, and Theraulaz 1999), "dumb power" (Kelly 1998), "spontaneous order" (Strogatz 2003, Regis 1994, Postrel 1998, Kelly 1993), and "hidden order" (Holland 1996) all play on a similar linguistic device: an oxymoron. These duos tie together the irrational (the fuzzy and undirected) with the rational (the instrumental, purposive, and focused). In all of them, a "bad" thing (which incidentally characterized Fordist society) is rendered "good" by the power of network technology and, more generally, by the architecture of the network. *Power*'s coerciveness and oppressiveness is curbed by being *distributed* in a way that flattens and diminishes the very poignancy of power; the threatening *mob*—a bundle of thoughtless individuals homogenized and manipulated by a "mass society"—becomes *smart* and thoughtful; and *order*, which we were led to believe requires centralization and control lest it devolves into entropy, is achieved *spontaneously*. This teasing use of oxymorons defies our intuitive and well-established notions of rationality.

With network technology, these idioms suggest, we are entering a new level of rationality that is superior to the old one both in process (which is rendered more democratic and collaborative) and in result (which becomes more instrumental and efficient). This type of superior rationality, as suggested by these duos, is inextricably tied to a new architecture of organization; rationality and network go hand in hand. These duos also do something else. They help imagine a notion of society comprised of individuals, and of social dynamics that are reducible to the unrestrained actions of free individuals. They suggest that the coordination of these individuals into a rational society comes about without any central, constructed, and overt mechanism but with one that is "hidden," "distributed," and "spontaneous." I will get back to this theme at more length later in this chapter.

Network technology is revolutionary according to the digital discourse since it provides the technological space for the leap from the irrational (nodes) to the rational (network) to take place.[3] The counterintuitive leap between the two orders—the simple-minded, short-sighted, dumb nodes on the one hand and the smart, rational collective, on the other hand—is contemplated also in "The Swarmbots Are Coming" (Dorigo 2004). The article begins with a conundrum from natural history: "Ants are simple creatures, yet they can perform complicated tasks," and, perhaps most surprisingly, do it "without any centralized control" (Dorigo 2004). How can the simple grow to be complicated? How can the dumb become smart? Most importantly, how can this be achieved without a centralized "brain"? The answer, the article suggests, is to be found in the architecture of the network, where smart mobs can thrive. Making an unproblematic and direct leap between the biological and social orders, the author suggests that ants can teach us how to make the economy work better: "Boil down ant behavior and what do you get? A new set of business tools known as ant algorithms: basic behaviors that can be programmed into a large number of independent software agents to solve human problems . . . The ability to swarm, adapt, and optimize makes ant algorithms a crucial technology for the information age, especially as everyday objects become ever smarter. The rules that insects live by turn out to be perfectly suited to the high tech anthill" (Dorigo 2004). Hence, order and rationality emerge spontaneously and effortlessly, not by some ingenious and centralized brain, but by the networking of multiple simple nodes.

These are the technological bases of Kelly's elaborate analysis of market rationality in the digital age. These characteristics of the network, says Kelly, are now leaking through the ubiquity of network technology into the workings of the economy. Economic rationality and economic growth have now come to be dependent on the harnessing of the power of the "distributed bottom" (Kelly 1998, 14), and the workings of the economy have increasingly come to resemble this technological form. Hence, when Kelly asserts the centrality of communication to the network economy, he is specifically excited about the possibility of a greater *market rationality* to emerge from the interconnection of small and simple parts. Simply put, the weaving of network technology into the fabric of the market renders its workings more rational.

ORDER WITHOUT ORDERING: DECENTRALIZATION AND DEHIERARCHIZATION

As stated previously, one of the most important implications of this new rationality is that it cannot be managed, programmed, and directed from any central point. It works best when it is left to its own devices—that is, when it is self-regulated. Luckily, the argument goes, the requirement for self-regulation can now be satisfactorily attained with network technology. These complex systems (such as the market) can regulate themselves through the internal mechanisms of networks: decentralization, flexibility, and adaptability. Network rationality is markedly different from the previous model of rationality that prevailed in industrial society. It is in fact a complete reversal of industrial rationality, which entailed a hierarchical and centralized organization of command and control. As Nicholas Negroponte—founder of the MIT Media Lab, author of *Being Digital* (1996), and a contributor to *Wired* magazine—puts it, network technology undermines authority, since in a network "nobody is in charge . . . The Net—a reliable system composed of loosely connected and imperfect parts that work because nobody is in control—shakes up all our centralist notions, and hierarchy goes away by example" (Negroponte 1997). Networks entail (and require) the collapse of any centralized oversight.

Kelly further exemplifies decentralization and dehierarchization with Cemex. This Mexican company, Kelly reports, is able to excel where others have failed—delivering a time-sensitive product like cement—by switching from rigid centralized planning, to a just-in-time technique, where drivers, not management, schedule deliveries in real time. With the aid of network technology the drivers form a smart mob: "The drivers formed a flock of trucks crisscrossing the town. If a contractor called in and order . . . the available truck closest to the site at the time would make the delivery" (Kelly 1998, 15). In this, as well as other cases, the lesson that Kelly draws is that "no central brain coordinates; the [result] comes from the swarm of minibrains" (15). For Kelly, this is a story about the toppling of the centralized brain in the economy, an overthrow of the prenetwork, bureaucratic organization (be it management or the state) that could presumably grasp the whole picture and hence be able to coordinate the work of disparate units. This old type of big brain, as opposed to the new "minibrains," is too slow and inadequate to manage the complexity of systems such as a cement company, let alone the market. Such systems are indeed so complex that no *one* can manage them. But with the power of network technology, they can now manage themselves through the connection of many dumb *ones* into a smart whole. As Kelly puts it, "Dumb parts, properly connected into a swarm"—that is, through the tools of network technology, "yield smart results" (13).

The toppling of hierarchy as a characteristic inherent to the new network rationality is reiterated in "The New Facts of Life" (Meyer 2004). In Chapter 7 I elaborate on the underlying cosmology of the digital discourse, which considers network technology to be compatible with the universal laws of nature. Here, the narrative of network technology as facilitating a social transformation in accordance with the universal laws of nature is applied to explain the transposition of spontaneous order and mob action from the biological to the social realm. Social constructs, the article asserts, act just as if they were living things: "Markets and power grids have much in common with plants and animals . . . It turns out that many of life's properties . . . show up in systems generally regarded as nonliving" (Meyer 2004). Among these "life's properties" (i.e., the unchanging laws of the universe) now showing up in human-made systems are, according to the article, emergence and self-organization: "Emergence describes the way unpredictable patterns arise from innumerable interactions between independent parts . . . Self-organization is a basic emergent behavior. Plants and animals assemble and regulate themselves independent of any hierarchy for planning or management" (Meyer 2004). The underlying narrative is quite straightforward: organizations modeled after a network architecture work best when they are ungoverned and left to self-regulate.

The article further makes this point with the example of unmanned aerial vehicles, or UAVs. Programmed with only simple rules, UAVs are able "to direct themselves better than any dispatcher could" (Meyer 2004). They are networked and self-taught: "If one is shot down over Afghanistan, all drones everywhere gain improved responses to that form of attack." As the article points out, this network principle has been known in nature for ages. The way UAVs work "is precisely how bacteria develop resistance to antibiotics, only faster" (Meyer 2004).

But these network characteristics are now showing up not only in nature and technology but also in society. "Oddly enough," the article continues, "our growing knowledge of life processes could have its biggest impact in the social sciences. Social systems, after all, are made up of interacting agents, i.e., people. When we become adept at applying these insights to the social sphere, we'll be able to run simulations that reveal, say, the conditions under which Iraq would reconstruct itself. At that point, the new science of life will help us not only live better, but live better together" (Meyer 2004). Society, in this piece, is rendered a particular case of the general template of networks, individuals are "interacting agents," and society progresses through self-regulated, self-taught processes. We humans, this narrative implies, have been investing too much energy in planning, managing, and controlling social processes in a centralized, top-down manner, when really the best results can be achieved by simply letting the "distributed bottom" (Kelly 1998, 14) communicate, coordinate, and swarm.

According to Kelly, this new type of rationality, based on decentralization and self-regulation, is so innovative, revolutionary, and unintuitive that for the uninitiated, the working of the network may sometimes have the *semblance* of irrationality. For example, "Your [electronic] mail may go to Timbuktu and back on its way across town. A centralized switching system would never direct messages in such a wasteful manner. But the inefficiencies of individual parts is overcome by the incredible reliability of the system as a whole" (16). The network may look irrational, wasteful, and chaotic, but Kelly reassures—as we will further see—that these are in fact the very symptoms of a healthy and vibrant network and represent a new, superior form of rationality.

If decentralization is the path to building a society based on spontaneous order, what is the role of centralization and hierarchy in the network society? Kelly is no anarchist, and he certainly does not go as far as calling for the outright toppling of *any* form of central planning and governance. But the network, he insists, needs only "minimal governance" (18). Kelly, therefore, suggests focusing on the new territory of networks, arguing that "at present, there is far more to be gained by pushing the boundaries of what can be done by the bottom than by focusing on what can be done at the top" (18). The potentialities of a top-centralized governance have been historically exhausted, and it is therefore no longer a viable path for change. Progress can only come from exploring and tapping into the "bottom" option.

CHAOS AND PROGRESS

So far, I have explored the underlying *technologistic* arguments of the digital discourse regarding the emergence of a new rationality, which is the result of the interconnection of disparate nodes. I have also pointed out some of the social implications of such rationality. But the full force of the argument—and from the point of view of this book, where the hegemonic dimension of the digital discourse comes into light—comes when *Wired* switches from discussing technological, biological, and small-scale organizational scenarios to a discussion of the network economy as a whole. After all, the incentive for the technological discussion in *Wired* is ultimately to unravel its social consequences. As Kelly puts it, "The internet model has many lessons for the new economy" (16). As I said above, in the digital discourse, network technology is seen not only as a material basis for contemporary economy but also as its metaphor.

COMPLEXITY AND INCALCULABILITY

According to Kelly, the "link[ing of] the distributed bottom" (14) in the economic sphere renders the network economy a system so complex that it yields nonlinear, and therefore unexpected, incalculable results. This is in contrast to the industrial economy. In the industrial age, Kelly writes, "success was in proportion to effort" (31), and the future was therefore easy to predict: "To imagine the future of an enterprise or innovation one needed only to extrapolate the current trends in a straight line" (31–32). This simple algorithm fell by the sidelines "with the advent of large-scale electronic media networks in the mid century" (32). These technological systems no longer work as if they were man-made systems—which would make them easier to control—but rather as if they were living things. "Everyday we see evidence of biological growth in technological systems. This is one of the marks of the network economy: that biology has taken root in technology. And this is one of the reasons why networks change everything" (32). And now that biological principles took hold of the economy, its workings become more volatile and chaotic and hence less predictable.

The new economic rationality, spontaneously emerging from the nature of technological networks, is presented in the digital discourse as a mirror image of the instrumental rationality that was the backbone of the modern economy. Industrial economic rationality, succinctly encapsulated by Max Weber at the beginning of the twentieth century, entailed calculability, control, and predictability. The digital discourse heralds the emergence of a new rationality that features precisely the opposite symptoms: unpredictability and incalculability.

I have mentioned above the role that the biological imagination takes in forming the digitalistic worldview: information and communication technologies are seen as biological-like systems, which in turn create biological-like social systems. Here Kelly explains the fluctuating, unstable, turbulent new economy as *inherent* to the nature of networks, by using insights from biological systems: "As networks have permeated our world, the economy has come to resemble an ecology of organisms, interlinked and coevolving, constantly in flux, deeply tangled, ever expending at its edges. As we know from recent ecological studies, *no balance exists in nature;* rather, as evolution proceeds, there is *perpetual disruption* as new species displace old, as natural biomes shift in their markup, and as organisms and environments transform each other" (108, emphasis mine).

The argument is that as technologically enabled networks become the central axis of social activity, the economy has come to resemble nature: both are evolving progressively and are in perpetual imbalance. In fact, Kelly ties together those two processes—chaos and progress—to account for the realities of the new economy. Using an evolutionary framework, Kelly proposes that economic progress comes about *through* constant flux and disruption. These are not by-products or side effects of economic rationality and growth but the motor thereof. "Harmony in nature," Kelly asserts, "is fleeting" (108), and so it is in the new network economy: "Companies come and go quickly, careers are patchworks of vocations, industries are indefinite groupings of fluctuating firms" (108). Kelly implicitly directs his arrows of critique not toward these new realities but rather toward the outdated language used to describe and explain them. Thus, Kelly's description should be read as offering us a new lexicon, where new definitions, more appropriate for the new economy, substitute for traditional, industrial-age terms: "Careers *are* patchworks of vocations" (108, emphasis mine)—not the construction of tenured, stable, secure, progressive, lifelong professional tracks; "industries *are* indefinite groupings of fluctuating firms" (108, emphasis mine)—not demarcated, stable units and businesses. To treat careers and businesses as stable, in Kelly's view, reflects an imposition of an anachronistic framework (entailing linearity, stability, predictability, and harmony) of a bygone era on the new digital reality. The new economy is a network economy, and "networks are immensely turbulent and uncertain" (111).

FLEXIBILITY AND ADAPTABILITY AS REACTIONS TO CHAOS

Chaos is not a disruption of an otherwise stable network; rather, it is one of its chief characteristics. Chaos, in fact, becomes the new *sine qua non* of the economic environment to which economic actors need to react with flexibility and adaptability. As the title of another *Wired* article proclaims "Chaos Is the Form" (Hughes 1994), a title that alludes to the turbulent, unexpected form of technological advance. In this realm, the article says, "things are not the way they seem." Chaos and unpredictability are the only constants in the network economy, so much so that the best sign that something goes *wrong* is if everything seems to be going *right*: "Anyone in a business who understands it completely is probably already failing." This, according to the article, is particularly true to businesses in the technological field, where the success and failure of technological innovations is completely unpredictable. The article gives an example of a "bulletin board system operator, whose BBS operates out of his basement, grossed more than US$5 million last year. That BBS will probably be gone next year, but the operator had recovered his investment in the first two weeks of operation." On the other hand, "a lot of smug businesses, organizations, and implementations built around technology are going to be bypassed in the next few years—so fast it won't even be funny. I'll bet the Internet as we know it will be passé in five years—just as the largest number of people are waking up to it and making investment decisions about it. They will soon look foolish" (Hughes 1994).

This environment—chaotic, unpredictable, and formless—is emblematic of technological progress to such a degree that it in fact defines the new form of technological and economic activity. As the article concludes, "Change is driving everything. Chaos is the form" (Hughes 1994). In the same vein, Kelly addresses the increasingly chaotic and unstable working arrangements that characterize the network economy. The trend whereby full-time, long-term careers within organizations are substituted by an increasingly unstable and chaotic employment environment is interpreted by Kelly through the notions of flexibility.

Flexibility, according to the digital discourse, is a key dynamic in the network society, specifically in the network market. Flexibility entails two interlocking sets of meanings: reactive and proactive. The reactive set of meanings sees flexibility as the prime mechanism to cope with the chaotic nature of network markets and to adapt to it. It is in essence a defense mechanism of individual nodes (workers, companies, states, and so forth) against the fluctuating, chaotic nature of network market. In the proactive set of meanings, flexibility is seen as beneficial from the point of view of individuals, allowing them more degrees of freedom in their professional and creative life and increasing personal choices and freedom. I treat the proactive meanings of flexibility in the digital discourse in Chapters 4 through 6. In what follows, I focus mostly on the reactive set of meanings.

Kelly uses the employment schemes of the entertainment industry as an example for a healthy adaptation to the chaotic market through flexibility. In the entertainment industry part-timers, subcontractors, outsourced workers, and freelancers—all "convene as one financial organization for the duration of the movie project, and then when the movie is done, the company disperses" (Kelly 1998, 111). And the workers, one might ask? According to Kelly "after the [movie] gets slotted to video, everybody just vanishes" (111). In what sounds like a fantasy of employers in the flexible economy, once workers do what needs to be done for the ad hoc project, these "Hollywood film ad-hocracies" (111) just vanish. Flexibility, in this case, entails the construction of workers as atomized, individualized nodes who are required to adapt to the dictates of a network economy.

Indeed, flexibility is the key to success in the new economy. Information-based businesses—in sectors such as high tech, knowledge, and service—"rely on speed and flexibility to survive in a self-made speedy and flexible environment" (111). Such industries, according to Kelly, can only be managed with a "flattened, atomized" (111) business model. That is, they require the atomization of workers and their flexible and temporary employment arrangements. Here the model of the network economy as the space where individual nodes swarm together in bursts of rationality can be looked at from the opposite angle. The rationality of the network requires that nodes (specifically, in this case, workers) remain as atomized as possible so that their overall levels of flexibility and adaptability can be augmented. What might look like insecurity and instability from the point of view of workers is seen as a positive trend from the point of view of the functioning of the system.

It is precisely in this context that Peter Drucker—one of America's foremost business gurus—compliments American workers for their flexibility and adaptability (incidentally uncovering the economic structure that underlies such "virtues"). Referring to deep measures of economic restructuring entailed by the transformation from industrial to information economy, Drucker says, "In Germany, you have a social compact of the social market economy. Therefore, to lay off people, even to shift them, is very difficult . . . In those countries [Germany and Japan] it's very much a social rather than an economic problem . . . In this country [the United States], the restructuring has caused amazingly few social problems because our labor force is so mobile, so adaptable. Our disorder is a great advantage" (Schwartz and Kelly 1996). Here again, adaptability to market demands is seen as an asset—something that American workers posses and on which a network economy thrives.

The digital discourse does not simply describe the chaotic nature of the digital economy but also insists on the inevitability and benevolence of chaos. It suggests that since chaos is part and parcel of the network economy, it would be worthless and even dangerous to react to it defensively and try to counter this economic reality with a political and social action. In this context, Kelly invokes the anti-Luddite maxim that perceives any opposition to technologically induced change as a reactionary and futile struggle waged by shortsighted people. Such reaction, Kelly explains, is a symptom of "future shock," a term Kelly borrows from futurist Alvin Toffler, who coined it in 1970 to describe what he considered to be a natural human response to accelerating technological change. Only now the "shock" is even greater, since "the network economy has moved from change to flux" (Kelly 1998, 109). It is no longer the same type of "domesticated," linear change we became familiar with during industrialism. Instead, the nature of change itself has been paradigmatically transformed. It is a change that accelerates and morphs into creative destruction, a change that "overturns the old ways of change" (115). In the face of such radical change, shock may be a reasonable personal reaction, but in no way, Kelly suggests, should it guide any social action; our reaction should ultimately be to adapt to the new reality.

This moral imperative is brought home in an article that compares the culture of technological change in the United States and Europe (Glenny 2001). Europe, the article argues, shows a lack of interest in the new network economy: "Change *is* happening. But Europe's innate conservatism has made it a recalcitrant host to the rapid transformation sweeping the globe." This innate conservatism, he contends, involves "veneration of the state, and obsession with national identities and borders, and . . . a suspicion of—not to say hostility toward—science and technology" (Glenny 2001). At the heart of the author's critique is Europe's unwillingness to surrender to what are, according to the digital discourse, the determining and inevitable effects of network technology on the economy—chaos, acceleration, unpredictability, and instability—and her insistence on mitigating them by recourse to political means—that is, through state actions.

The Europeans, he says, tend to examine past "episodes of destruction [brought about by technology] that could easily be repeated." Their American counterparts, by contrast, are future-oriented and optimistic (Glenny 2001). But what does it mean to be future oriented and optimistic? To put the question in Toffler's terms, if, after the initial shock, we should adapt to the future, what is it that we should adapt to? On this question, Kelly cannot be more unequivocal. The economic instability and uncertainty that we now experience, he says, is here to stay. What might have been called in the past "change" has turned by now into "flux." And flux, unlike change, is no longer a road to stability, a difficult phase, or a bump *en route* to a solid ground but a permanent reality. Flux, in Kelly's words, is a constant state of "destruction and genesis. Flux topples the incumbent and creates a platform for more innovation and birth. This dynamic state might be thought of as 'compounded rebirth.' And its genesis hovers on the edge of chaos" (Kelly 1998, 109).

CHAOS AS A CREATIVE FORCE IN THE ECONOMY

But flux is not simply a new force to be reckoned with, an inconvenient "bad" we must now adapt to in order to enjoy the "good"; instead, it is a positive thing in itself. Hence, not only should flux not be tampered with or mitigated, but also, more than that, it should be encouraged. Thus, for example, instead of lamenting the loss of job security in the new economy, Kelly suggests we simply revoke our perception of what jobs are, "rather than considering jobs as a fixed sum to be protected and augmented . . . the state should focus on encouraging economic churning—on continually recreating the state's economy" (109). By employing outdated categories of economic life (such as predictability, calculability, security, and stability), Kelly implies, we might end up misjudging the benefits of a new network rationality. Rather than resorting to an industrial-age framework to evaluate the present, Kelly suggests we should come to grips with the chaotic and fluctuating characteristics of networks.

More than that, chaos and flux should be actively sought and encouraged. Taking his cue from nature again, Kelly reports what ecologists, familiar with the notion of constant flux, have learned: "The sustained vitality of a complex network requires that the net keeps provoking itself out of balance" (110). Rather than seeking stability and certainty of the economy, and rather than attempting to work toward harmony and balance, network societies should encourage and provoke conditions of flux and chaos in their economy. Flux is quintessential to the health of the network economy; the network economy, he says, thrives on its own destruction, and our goal should therefore be "to sustain a perpetual disequilibrium" (110) rather than to mitigate it. He wraps up this lesson with the Stalinist-sounding slogan: "constant innovation is perpetual disruption" (110).

As I have argued in the Introduction, the digital discourse offers not just a *new* worldview that replaces an old-fashioned one, but one that constitutes a *critique* of Fordist definitions of economic life. This is succinctly exemplified in the "Encyclopedia of the New Economy" (Browning and Reiss 1998). Chaos, according to the entry in the encyclopedia, means not disorder but quite the opposite. It is order without ordering, order without central oversight, without planning, but an order nonetheless, order that does not follow "a strict rule book," which characterized the top-down, centralized, planned industrial economy, but that represents "an evolving collection of shared goals" (Browning and Reiss 1998).

Lastly, we have already seen the careful construction of the technologistic argument, which explains why the network economy—a decentralized, self-regulating system built on disparate and atomized nodes, working together through mechanisms of spontaneous order—is inherently chaotic and in constant flux, demanding flexibility and adaptability on the part of nodes. But even if one accepts this premise, another question remains to be answered: Why should we accept and even encourage such flux if it leads to a constant state of uncertainty and even "hovers on the edge of chaos" (Kelly 1998, 109)? Why, in other words, should we not control and mitigate it?

The answer, according to the digital discourse, is that this chaos is at the heart of the most important factor of economic growth in contemporary society: technological innovation. Chaos is *both* a breeding ground for technological innovation and the product thereof. Flux is a precondition for technological innovation: "Innovation," says Kelly, "is the productive and desirable moment between ordinary and insignificant change on the one hand, and a change too radical to be implemented on the other hand"; it is located on the borderline between "the rigid death of planned order and the degeneration of chaos" (113). To foster technological innovation, the motor of economic growth in the network society, we need to willingly occupy the space at the edge of planning and order, the space at the intersection of complete disorder. This place is precisely the network: "The ideal environment for cultivating the unknown," Kelly says, "is to nurture the supreme agility and nimbleness of networks" (Kelly 1997). The "trick" to fostering innovation, according to Kelly, is to have an environment favorable of change with as little paralyzing rules as possible and as little friction as possible. Rather than wanting to mitigate and curb flux and chaos, we need to accept that "the price of progressive change in maximum doses is a dangerous (and thrilling) ride to the edge of disruption" (Kelly 1998, 114). Hence, technological innovation, the new dynamo of economic and indeed human progress, makes the network economy inherently chaotic. Chaos and progress are inextricably tied.[4]

The digital discourse expects that the network economy will be much more turbulent than the industrial economy. What is left out of this representation of reality is the fact that the stability and predictability of industrial economy was not merely the product of its different nature, as implied by the digital discourse, but precisely a product of the political and social barriers put forth by governments on markets. Stability through the curbing of flux and chaos was precisely what governments tried to achieve through the construction of social democracies. The welfare state, the New Deal, Keynesianism, corporatism, embedded liberalism—all were varied attempts to reduce the instability associated with *laissez faire* economics and to provide at least minimal protection to individuals against unpredictable markets.

It is precisely against this historical backdrop that Kelly makes a revealing statement regarding the underlying *political* (rather than merely technological) project entailed in the construction of the new economy: "In a poetic sense," Kelly says, "the prime goal of the new economy is to undo—company by company, industry by industry—the industrial economy" (112). It is precisely the *undoing* of the political constraints put on markets and the layer of social arrangements that were constructed throughout the twentieth century in order to insulate individuals from an unforgiving, unpredictable, and irrational markets (in the broader sense of substantive rationality) that the digital discourse is calling for and legitimates through a naturalization and technologization of flux and chaos. What the digital discourse constructs is a structure of the network market as a chaotic but self-regulating mechanism, the operation of which is best achieved by leaving it to its own devices and in no way regulating it politically.

SPONTANEOUS ORDER AND CHAOS IN NEOLIBERAL THEORY

The digital discourse presumes to decipher contemporary society in terms of the revolution of network technology. Dumb nodes, spontaneous order, smart mobs, decentralization, flexibility, adaptability, chaos, and flux—all these underlying characteristics (or rules, in Kelly's terms) of the network economy are presumed to be predicated on and flow from the dominant technological form. In this sense, these characteristics are also presumed to be a technical, neutral, and apolitical representation of the network market. But as we delve deeper into the digital discourse's representation of the network society, another source emerges as its conceptual toolkit. The digital discourse on the network market has an uncanny resemblance to the dominant political culture of post-Fordism: neoliberal theory.

This part of the chapter explores this resemblance, highlighting both similarities and differences between the digital discourse and the neoliberal discourse. The purpose of such comparison is to delineate in clearer terms the political sources, or the "political unconscious" (Jameson 1982) of the digital discourse on the network market. In other words, it points to the homology of the digital discourse's *technologistic* discourse and neoliberalism's *economic* discourse.

Spontaneous Order

The scope of neoliberal theory is wide, from its roots in neoclassical economics (Adam Smith) and liberal thought (David Hume), through its reincarnation as neoliberalism in the 1960s (Friedrich Hayek and Milton Friedman), and its political implementation as Thatcherism and Reaganism, to its current popular articulation in the context of globalization (Thomas Friedman and Francis Fukuyama). At the risk of oversimplification, let me present some of the core tenets of neoliberalism through the concepts of spontaneous order and chaos, especially as these are developed in the work of Friedrich Hayek, arguably the most distinguished figure of this school. More forcefully than other key concepts, spontaneous order and chaos seem to bridge neoliberal theory and the digital discourse and can serve as points of entry for our discussion here of the mediation of neoliberal theory into digital discourse in particular and the mediation of politics through technology in general.[5]

Spontaneous order is arguably the single most important theoretical concept in neoliberal theory (Petsoulas 2001; Sally 1998). Neoliberalism argues that, perhaps contrary to our intuition, social order is not necessarily a result of a conscious, planned design, but springs spontaneously. An exemplary case in point, and indeed the epitome of all social orders, is the free market. There is no directing hand designing the market, but order nevertheless comes about through the interaction of disparate units. Each of these units—in the case of markets, individuals—follows its own selfish and narrow rationale and adheres to its own interests. According to this theory, markets are "developed spontaneously by individuals as a mechanism for the efficient satisfaction of their desires" (Ashford and Davies 1991, 170–73). But in the aggregate, the multiplicity of selfish and disparate actions results in an overall order, which is socially rational and benevolent. Spontaneous order, and specifically the market, is superior to any human-planned order for a number of reasons. It is universally rational and beneficial, apolitical, and, most significantly, it is a self-regulating mechanism. In fact, attempts to regulate or plan the market are likely to interfere with its self-regulating, spontaneous mechanisms and cause more damage than help. Therefore, neoliberal theory strongly advises that markets be insulated from the interference of planned and centralized orders, such as the state (with its regulations, legislation, and so forth), as well as civic organizations such as trade unions.

The central narratives of the digital discourse regarding technologically enabled networks as examples of spontaneous order are nearly parallel to the neoliberal treatment of markets. Likewise, the digital discourse's advocacy for the superior rationality of the network market is very much akin to Hayek's advocacy for the superior rationality of the free market. The genius of the market, according to Hayek, is anchored in the characteristics of spontaneous order. The principle idea behind the theory of spontaneous order "is that society and its institutions are neither 'natural' formations nor the outcome of human design; instead, they originate in the unintended and unforeseen spontaneous coordination of a multiplicity of actions by self-interested individuals" (Petsoulas 2001, 2).

As in the digital discourse's notion of networks, in Hayek's notion of markets, too, rationality emerges out of irrationality. Rationality is both unintended and unforeseen; it is impossible to predict much less to design and direct. Rationality and order, however, are possible to be recognized by evaluating the pragmatic results of spontaneous formations: "A spontaneous formation can be characterized as *order* when it has a *structure*, and also when it is *beneficial* for the individuals involved" (12). At the heart of both networks and markets is not a conscious effort to design order according to plan but simply the unforeseen beneficial outcome of the coordination of multiple and disparate actors.

The digital discourse does not simply reiterate neoliberal tenets but digitize them, so to speak—that is, render them technological. Thus, the comparison of the two discourses can be likened to laying two nearly identical slides on top of each other, thereby revealing both similarities and differences. This is evident, for example, in the discussion of the status of individual nodes *vis-à-vis* the network. In the digital discourse, individual nodes are perceived as inherently inferior in rationality and smartness compared with those of the network as a whole. It is only through the decentralized, self-regulated interaction of "dumb nodes" within a network that superior rationality can emerge. In the digital discourse spontaneous order is inextricably linked with the limited rationality of nodes *vis-à-vis* the rationality of the network as a whole.

These ideas receive a similar articulation in neoliberalism, where markets and ignorant individuals substitute for technological networks and dumb nodes, respectively. However, in the absence of technologistic argumentation, the underlying explanations for the same surface assertions are somewhat different in neoliberalism. According to Hayek, the key reason that the market ends up yielding the most rational results stems from the status of knowledge and information in society. Like computer chips or nodes in networks, so is each individual in society—dumb; more eloquently, Hayek refers to the "constitutional limitations of the human mind" (Petsoulas 2001) or to individual's "constitutional ignorance" (Sally 1998, 19): "Knowledge of relevant facts for market exchange is highly fragmented and highly dispersed among millions of individuals in complex societies" (19). No *one* person has the capacity to muster the complete knowledge required to come up with an intelligent design of the economy. Spontaneous order is the means by which dispersed and tacit knowledge, fragmented among millions of individuals, is processed and rendered so that it yields the most rational results: "Compared to designed orders, spontaneous social orders are more complex, and utilize knowledge which no single individual or even a group of individual minds would ever be able to grasp, let alone control. Such knowledge would be impossible to centralize because it is of a practical nature, depending on the particular circumstances in which individual participants find themselves" (Petsoulas 2001, 2). For Hayek, the problem of fragmented and partial knowledge of each individual in the marketplace—the "epistemic problem" in his words—is answered best by reliance on "the spontaneous ordering forces of the market" (2).

Thus, that the nodes comprising the network—be it computer chips, workers in a company, companies in the economy, or individuals in the marketplace—are dumb, unreflexive, and short sighted relative to the network as a whole is a cornerstone of both the digital and neoliberal worldviews. In the digital discourse, it is premised on a techno-scientific discovery of the operation of nodes in information networks. In neoliberalism, it is premised on the socioscientific assumption regarding the limited capacity of individuals: individuals are "partially and perpetually ignorant" about markets (Sally 1998, 19). The crucial difference between the digital discourse and neoliberal theory is the depth in which the foundations on their respective arguments lie: while in neoliberal thought these are anchored in abstract, theoretical constructs such as "the market," or "constitutional ignorance" (i.e., in knowledge), in the digital discourse these arguments are welded into the "materiality" of network technology, such as dumb chips and network architecture (i.e., in practice or reality).

Both in the digital discourse and neoliberal theory, the constitution of the individual as the central and sole unit of social operation is paramount. Both place premiums on the independence of each node in the network (or liberty of each individual in the market). But there is also an interesting variance between them, a variance that might be conceived in terms of an "evolution" in the analysis of the interrelation between individual liberty and spontaneous order from classical liberalism, through neoliberalism, and finally to the digital discourse. In the classical liberalism of Adam Smith, John Stuart Mill, and David Hume, individual liberty in the marketplace is seen as a natural, unconditional right, hence the term "natural liberty"; it is assumed to be a virtue on its own merit (Greenwald 1994; Ashford and Davies 1991, 170–73). This constitutes a *normative legitimation* for the free market.

In contrast to these scholars of the Scottish Enlightenment, neoliberal theory withers away with the normative and metaphysical component of "natural liberty" as a basis for liberalism. Neoliberalism turns the argument on its head, seeing individual liberty as a prerequisite for the successful operation of the market. For spontaneous order (i.e., the market) to occur, individual liberty must prevail. For Hayek, the argument for liberalism is based on a social theory rather than on a moral premise. His advocacy for individual liberty stems from the "epistemic problem" discussed previously (Petsoulas 2001, 2, 16): "Given that knowledge is dispersed, tacit and temporary, it is most efficiently utilised in an environment of decentralised decision-making in which *individuals are free to pursue their goals* by using the information available to them" (16, emphasis mine). This constitutes a *scientific legitimation* for the free market. As

Hayek comments in a related matter, liberalism "derives from the *discovery* of a self-generating or spontaneous order in social affairs" (Hayek 1967, 162, cited in Kley 1994, 1, emphasis mine); that is, it is grounded in science and facts.

This claim for the liberty of individuals has also been articulated in Hayek's social theory around the concept of "methodological individualism," which suggests that "all actions are performed by individuals; therefore analysis of social reality must start from individuals." Moreover, it suggests that "a social collective has no existence and no reality beyond the actions of its individual members" (Gamble 1996, 53). This atomized conception of society, where society is "nothing more than a collection of individuals" (Rapley 2004, 77), has also been popularized by a slogan coined by Margaret Thatcher—a central protagonist in putting neoliberal theory into practice—who famously announced that there was "no such thing as society, only individual men and women" (cited in Harvey 2005, 23).

The digital discourse follows much of the neoliberal, scientific legitimation, especially in arguing that spontaneous order is a predominant phenomenon in nature. But it also transforms the argument for individual liberty into a *technologistic legitimation*. To the extent that individuals are reconceived as nodes within a technological network, and to the extent that these nodes must be atomized, flexible, and adaptable to network fluctuations, individuals must be free in order for spontaneous order to occur. Hence, the argument for the liberty of individuals in the context of free markets is asserted thrice: first, on normative grounds, second, as a scientific discovery, and, finally, as a technological necessity.

Like the digital discourse, neoliberal theory, too, is concerned with explaining how market rationality emerges from what might be seen as haphazard, disorganized, individualistic, ungoverned, and even conflicting actions. In neoliberal theory, spontaneous order mediates individual "micromotives" into "macrobehavior" (Sally 1998, 1) and "private vices" into "public benefits," in Mandeville's terms (see Petsoulas 2001, chap. 3). Spontaneous order accounts for what might seem an unbridgeable gap between irrationality and rationality, between multiple particularities and universality. According to Hayek, spontaneous order does the trick by providing the best tool for the allocation of knowledge; it is the best solution for individuals' epistemic limitation. Sally explains, "The genius of the market economy is that it allows individuals to use freely what partial knowledge they have in everyday activities of production and consumption. The competitive market is by a long shot the best available device to coordinate existing (fragmented, dispersed and tacit) knowledge, as well as to create new knowledge, in order to cater for material wants" (Sally 1998, 19).

In a famous passage from *The Wealth of Nations*, Adam Smith, too, grapples with the quantum leap from unreflexive, or dumb micromotives to a rational, beneficent macrobehavior:

> Every individual necessarily labours to render the annual revenue of the society as great as he can. He generally, indeed, neither intends to promote the public interest, nor knows how much he is promoting it. By referring the support of domestic to that of foreign industry, he intends only his own security; and by directing that industry in such a manner as its produce may be of the greatest value, he intends only his own gain, and he is, in this, as in many other cases, led by an invisible hand to promote an end which was no part of his intention. (Smith 1776, 32)

"Spontaneous order" and the "invisible hand" are metaphors for a coordinating mechanism of the activities of individuals who, apart from their own production ability and consumption needs, know very little of the wider economy and who only have the satisfaction of their own interests in mind (Sally 1998, 19). As does the digital discourse, so neoliberalism posits the transformation of dumbness and selfishness into smartness and social benevolence. And both the digital discourse and neoliberal thought are able to conceive of a society, which is nevertheless reducible to its individual units: a network of nodes and a market of individuals.

CHAOS

So far I have compared the digital discourse and neoliberal theory in their respective discussions of networks and markets, especially through the notion of spontaneous order, central to both these worldviews. I have already presented the digital discourse's discussion of the chaotic nature of network economy, as well as the delimitation of social action mostly to adaptation and flexibility on the part of individuals and *laissez faire* policies on the part of society as a whole. The resemblance of this digitalistic representation of the economy to neoliberal theory is again uncanny.

The notion of chaos is also central to neoliberal understanding of markets. Here, again, it is worthwhile noting the transformation from (neoclassical) liberal thought, through neoliberal discourse, and finally to the digital discourse in terms of accounting for the realities of markets. A crucial distinction between Adam Smith's notion of the invisible hand and Hayek's notion of spontaneous order is their teleology. For Smith, the invisible hand brings harmony and homeostasis; markets incline toward equilibrium. In contrast, for Hayek, the natural state of the economy is disharmony. The market is not seen as a mechanism designed to attain homeostasis through equilibrium of supply and demand. Instead, the market "is a dynamic *process* of endogenous change and growth in which means, wants and technology are constantly transformed and in which individual activity is exploratory and experimental. To Hayek, competition is less about equilibrium and more about a 'discovery procedure,' an ongoing, open-ended process that coordinates and generates knowledge in a decentralised manner to adapt to uncertainty and environmental flux" (20).

Hence, Hayek's theory is a critique of the neoclassical equilibrium theory. Markets are "always in disequilibrium" (Ashford and Davies 1991, 170–73) and in "flux" (Sally 1998, 20), fostering "restless individuals" to engage in a perpetual "discovery procedure" (Ashford and Davies 1991, 170–73). Both the digital discourse and neoliberal theory expect spontaneous order to be in perpetual flux and both recommend the same recipes to cope with that: for the individual—adaptation through flexibility; for states—*laissez faire* policies and withdrawal from the market (and, by proxy, from society).

There is a straight line going from the digital discourse's ideas about the chaos of markets and the connection between chaos and technological innovation and progress to the ideas of another prominent figure in the school of Austrian economics (11–14) and a contemporary of Hayek.[6] For Joseph Schumpeter, too, capitalism is inherently and benevolently marked by a state of disequilibrium. Schumpeter assigns a prime role in a capitalist market to technological innovations. These innovations bring about a state of "creative destruction": they create new wants and products and, at the same time, render older wants, commodities, and jobs obsolete. Therefore, "creative destruction" is a prime motor of economic growth, but, at the same time, it makes capitalist markets unpredictable, combining "long-term growth with boom-and-bust cycles" (Fuller 1999). Periods of great economic turbulence are therefore seen as an indication that the market is in the midst of a (benevolent, positive, and progressive) process of *creative* destruction. The state is advised not to interfere in the market in such times and not to mitigate these great disruptions.

This last point by Schumpeter brings us to the final parallel between the digital discourse and neoliberal theory: a strong advocacy for the insulation of the market from the democratic political process. As we have seen above, in the digital discourse it is information networks that render the operation of markets more rational. More precisely, the rationality predicated on the spontaneous order that emerges from the decentralized coordination of disparate nodes materializes and reaches its full potential with the digitization of these procedures. The rationality of network markets is suggested to be, in the digital discourse, universal and apolitical.

Not necessarily for the same underlying reasons, neoliberal theory, too, makes a case for the insulation of markets from political interference (Rapley 2004, 76). The insulation of markets from politics is premised on two arguments: that planned order is inferior to spontaneous order and that political intervention hurts the mechanism of self-regulation. Let us take a closer look at each of them. According to neoliberal theory, spontaneous order (or "cosmos," in Hayek's terminology) is superior to made order (taxis) on two accounts:

- *Complexity*: Given the limitations of the human mind, made order is bound to be simple, Cosmos, however, can be much more complex, involving many more variables. The inability of individuals "to grasp the complexity of social order and understand the forces that shape it, set 'limits to the extent to which conscious direction can improve upon the results of unconscious social processes'" (Petsoulas 2001, 4).
- *Purpose*: Made orders are constructed to satisfy the purpose of their makers. Spontaneous orders do not serve any particular interest. They have a function, rather than a purpose, which suggests awareness and interests (Petsoulas 2001, 13–14).

Thus, according to neoliberal theory, spontaneous order, specifically the market, is inherently apolitical in two distinct meanings of the term. First, given the complexity of variables and knowledge entailed in the construction of markets, it cannot be subjected to political processes; its complexity is so immense as to make the realm of politics ill suited to handle it. And second, markets are apolitical because their spontaneous emergence renders them cleansed of particularistic interests. They are seen as neutral tools that perform a disinterested function.

This is the crux of neoliberal conservatism, a conservatism that is not merely disguised but outright denied by the digital discourse. Rationality is *already* embedded in social institutions (those that are the result of spontaneous order). As Hayek puts it, "Our social existence is the product of evolutionary forces that are beyond any individual's capacity to comprehend fully. Traditional (evolved) practices and institutions embody the cumulative knowledge of past generations" (Petsoulas 2001, 4). Institutions and morals, such as private property, private law, money, and competition are "the result of human action, but not the execution of any human design"; they are "unintended by-products . . . of human action" (Sally 1998, 22). If we try to introduce planned order we soon find out that, compared to the merits of competitive markets, "central planning, and even *ad hoc* measures of government intervention, are much inferior in allocating goods and services. Governments lack access to and control of requisite information in order to plan or guide markets, and what little information they marshal is coordinated in a centralised and cumbersome, not to say ham-fisted, manner" (19–20).

Hayek, therefore, advises us that "as individuals we should bow to forces and obey principles which we cannot hope fully to understand, yet on which the advance and even the preservation of civilization depend" (Hayek 1979, 162, cited in Petsoulas 2001, 4). The social, thus, is not and should not be a product of conscious and purposeful—that is, political—construction. Society and culture do change, but this change should be left to internal, self-regulated, evolutionary processes and not be led by conscious, rational, and deliberate attempts. "We cannot redesign," Hayek says, "but only further evolve what we do not fully comprehend" (Hayek 1981, 167, cited in Petsoulas 2001, 4–5). Hence, all that humans can do is act in the most immediate, bottom-up, unreflexive, untheoretical fashion as atomized nodes in the network.

Neoliberalism strongly opposes any type of supposedly external regulation, one that has not organically and spontaneously spawned from the rationale and workings of the market. This refers not only to the state but also to organizations in civil society, such as labor unions. The only role appropriate for governments is to foster an environment that allows markets to function properly. That is, governments should protect lives, define property rights, and enforce private contracts; in other words, governments should act as enforcers and protectors of markets (Greenwald 1994).

Intervention of states in the market in fact hurts the mechanisms of self-regulation: "The plethora of *ad hoc* interventions by governments in the process of resource allocation . . . abrades and even destroys the self-generating and self-organising properties of the spontaneous market order" (Sally 1998, 23). "Political considerations" are therefore said to "typically interfere with economic logic" (Rapley 2004, 76). Indeed, somewhat euphemistically, market failures—such as phases of depression or high rates of unemployment—are commonly referred to as "government failure" (Yergin and Stanislaw 1998). As Ashford and Davies put it, in neoliberal theory, "The existence of 'market failure' can frequently be located in the creation of obstacles by government" (1991).

CONCLUSION: FROM SAFETY NET TO THE INTERNET

This chapter analyzed the digital discourse concerning the interweaving of network technology with the market. According to the digital discourse, network technology transforms markets in two fundamental ways. At all levels, from organizations, through industries, to the global economy, markets have become decentralized, dehierarchized, and flexible. The reconstitution of markets in accordance with the architecture of networks has rendered them more conducive to spontaneous order. Market order no longer has to be planned *a priori* by conscious decision and implemented top down; instead this organization is shown to increasingly emerge bottom up from the spontaneous actions of dumb nodes. Likewise, spontaneous order does away with the need for most forms of regulation and planning. Moreover, while network markets require less planning, intervention, and governing coordination, they nevertheless yield more rational results. Spontaneous order is predicated on, and in turn furthers, a new balance of power between individuals and society: a network market empowers individuals at the expense of social regulation through the state.

But top-down management—of the private company or national economy—becomes not only unnecessary but also virtually unfeasible because network markets are also more chaotic. This is the second fundamental transformation entailed by the rise of the network market. Market rationality does not simply increase quantitatively, but rather it changes qualitatively, featuring more flux, unpredictability, acceleration, change, and perpetual instability. This new economic reality requires individual actors, or nodes in the network, to react to the ever-changing market environment with flexibility and adaptability.

The digital discourse on the network market represents a fundamental shift in the political culture of contemporary capitalist societies from social democracy to neoliberalism or from embedded markets to market fundamentalism (Somers and Block 2005). It is against this background that the digital discourse should be understood. It is part of the new spirit of capitalism that sees the contemporary phase of capitalist society as an overcoming of the pitfalls of Fordist society and as the embodiment of the humanist critique of capitalism (Boltanski and Chiapello 2005)—specifically, regarding the central planning and organization of the economy by society and the state. Marketization and disorganization are seen in the digital discourse as stemming from network technology, as benevolent, and as a transcendence of the shortcomings of the Keynesian welfare state. As this book argues, as it promises the increased rationality of the economic system as a whole and more individual empowerment, the digital discourse also links these benefits to the precariousness and risks associated with the dynamics of a self-regulating market, such as flux and flexibility. It thereby works ideologically by depoliticizing, naturalizing, and neutralizing these social transformations.

MARKETS AND TECHNOLOGY LEGITIMIZATION

The rhetorical affinity of neoliberalism as a market ideology with contemporary discourse on technology has been well documented before (see, for example, Frank 2000; Barbrook and Cameron 1996; Best and Kellner 2000; Gere 2002; Borsook 2000; Turner 2006; Aune 2001; Mosco 2004; Wajcman 2004; Harvey 2005; Dean 2002). In *One Nation Under God*, Thomas Frank (2000) identifies the discourse on information technology as one of the key factors in popularizing market ideology. Books such as Walter Wriston's *The Twilight of Sovereignty* and George Gilder's *Microcosm* made the argument that information technology made the restrained form of capitalism (social democracy) obsolete and a return to nineteenth century-style *laissez faire* inevitable (Frank 2000, 54–55). Information technology came to be "the most powerful symbolic weapon in the arsenal of market populism" (57). Frank concludes, "Since the moment the Internet was noticed by the mainstream media in 1995, it has filled a single and exclusive position in political economy: a sort of cosmic affirmation of the principles of market populism" (79). As another author puts it, the discourse on information technology played a decisive role in *Selling the Free Market* (see Aune 2001, chap. 7) during the 1980s and 1990s. Moreover, Frank points out the transposition of market enthusiasm into a technological language. No longer was this enthusiasm bluntly ideological but it became technical: "Now the ideology seemed to emerge as a natural consequence of the technology being discussed rather than from the random floating anger of betrayed patriots" (Frank 2000, 79–80).

In the same vein, Barbrook and Cameron (1996) christen the conflation between information technology and market ideology the "Californian Ideology." In this techno-political vision, they say, the convergence of information and communication technologies is seen as leading to "the *apotheosis of the market*—an electronic exchange within which everybody can become a free trader" (Barbrook and Cameron 1996). According to this vision, network technology embodies an ideal of the free market (Robins and Webster 1999, 67). The Californian Ideology presents not only a new vision *for* society but also a new vision of *what* society is. Rather than seeing society in terms of structures and institutions, it sees information society as a network of free-floating individuals, who meet in the market place in order to trade and exchange ideas. According to the Californian Ideology, information technologies *inherently* "empower the individual, enhance personal freedom and radically reduce the power of the nation-state" (Barbrook and Cameron 1996). The fact that these outcomes are inherent to the technology makes any intervention of regulatory bodies (most notability, governments) an anachronism that is doomed to fail.

What these analyses share in common is a perception of the digital discourse as an ideology in the Marxist sense: an ideational construction that conceals material reality. Such an approach is also articulated in the work of Best and Kellner (2000), who criticize Kelly's analysis of contemporary society for ignoring the realities of capitalism that still prevail. They limit their discussion largely to refuting Kelly's arguments about the network economy by upholding the centrality of capitalism in shaping contemporary society. The ideological thrust of Kelly's discourse, according to Best and Kellner, is anchored in the biological framework he is using in order to provide a social analysis. Kelly, they contend, collapses the dividing lines between biology and society and transplants the new model of complexity theory from the natural world to the social world. They reject this unproblematic extrapolation of complexity theory from nature to society and see this blurring between nature, technology, and society as mystifying and depoliticizing the restructuring of capitalism along neoliberal lines by resorting to a language of inevitability.

Such analysis presents the digital discourse as a concealment of the new realities of capitalism. The thrust of the analysis offered in this chapter is different inasmuch as it situates the digital discourse on the network market in its historical context and interprets it within the analytical framework of legitimation discourse. According to this analysis it is not so much that the vector of capitalism is *externalized* from the digital discourse; instead, the realities of the new capitalism are very much *internalized* within the discourse, but they are masked and given technological clothing. Put differently, the digital discourse both articulates and legitimizes the transformations of capitalism.

Somers and Block (2005) use the term "ideational embeddedness" to account for the relations between an ideology of market fundamentalism and policies that have direct economic and social effects; market practices are embedded within a broader set of ideas and ideologies, which, they say after Bourdieu, have the power to create what they purport to describe (Somers and Block 2005). Following Somers and Block, I see the digital discourse as providing the ideational embeddedness for the new realities of capitalism and its new spirit. To put it somewhat ironically, the approach I take is to accept the digital discourse as a *true representation*, albeit, a true representation of a *different* object than the one the discourse purports to represent. In other words, the digital discourse represents not merely a new techno-social reality but also a new dominant ideology, political culture, and spirit that accompanies and supports it.

The comparison of the digital discourse to the neoliberal discourse sought to go beyond their overt ideological affinity (promarket and antigovernment) and explore the rationalizations and theorizations that underlie these assertions. What we have seen is that the digital discourse does not simply reiterate neoliberal tenets but translates many of the neoliberal tropes into a digitalistic language, rendering the deeper theoretical claims of neoliberalism digital. In that sense, the distinctions between the two discourses are no less revealing than the similarities, since most of these dissimilarities are a result of the digitalization of the tenets of neoliberalism. This is perhaps epitomized in the notions of markets and networks, as they are used in the digital and neoliberal theory, respectively. The market is an abstract construct, a scientific discovery, a social fact, in Durkheim's terminology. Networks (as they are construed by the digital discourse), on the other hand, stem from, and are anchored in, a material reality: the web of information and communication technology spanning virtually all geographical and social space. In that manner, *a priori* intellectual assumptions put forth by neoliberal theory are reaffirmed *a posteriori* by technological evidence in the digital discourse.

The significance of the digital discourse lies not in its overt *embrace* of free market ideology (as Barbrook and Cameron 1996, for example, point out) but—to use a somewhat harsh rhetoric—precisely in its *rejection* of ideology as such. The digital discourse strives to be precisely what a free market ideology, like neoliberalism, might have a hard time being—not an ideology at all. Unlike neoliberalism, it is based not on intellectual ideas, cognitive constructs, and abstract metaphors and models, and it has no overt political trajectory. Instead, it builds its foundations on the seemingly technical, materialist, and instrumentalist reading of technology. It is this "technological hermeneutics" that gives it a gloss of an impartial, apolitical rendering of reality.

As an analytical framework to explain and legitimize the realities of free market—that is, as *ideology*—the digital discourse therefore seems superior to neoliberal theory because it anchors much of the neoliberal arguments in material tools. If rationality is a product of the disparate and selfish wants of individuals, if it emerges spontaneously and is self-regulating, and if it requires a mechanism of communication, then the market, once digitized and incorporated into cyberspace, promises to be the most sophisticated market in the history of humanity. In that sense, the digital discourse not only reiterates but also *supersedes* the neoliberal arguments regarding the operation of markets by embodying it within network technology. In the digital discourse, economic rationality is redefined as emanating solely from the operation of networks, and so it is inextricably bounded with network technology.

It is therefore no surprise that advocates of neoliberalism are enthusiastic about network technology no less than technological enthusiasts seem excited about neoliberal ideas (Gere 2002, 140–41; Harvey 2005,

3–34, 157–59). In neoliberal theory the market is seen metaphorically as a machine for the coordination of the interests and actions of free individuals in a rational benevolent fashion. In the digital discourse, and with the introduction of network technology, this machine is no longer merely a metaphor; it is a reality, assumed to reaffirm and fortify the neoliberal claims. Thus, Thomas Friedman—who perhaps more than anyone in the public sphere epitomizes the synergy of network technology and neoliberalism—can write, "The Internet offers the closest thing to a perfectly competitive market in the world today" (Friedman 2000, 81). And Milton Friedman, the most prominent figure in neoliberal thinking in America, recently made similar assertions, noting that "the Internet . . . moves us closer to 'perfect information' on markets" (Friedman 2006). By percolating through a technologistic framework, the postulates of neoliberalism are added a gloss of realism by which they are vindicated and affirmed not only intellectually but also technologically.

   That technology seems to vindicate a particular political and economic system is of course not unprecedented. In a work of history, Otto Mayr examines the discourse on the technology of the mechanical clock in seventeenth- through nineteenth-century Europe (Mayr 1986). He shows the role that this technology played as a metaphor in the political discourse of the day. What is most striking about his account is that the clock came to serve as a metaphor for two *conflicting* political doctrines. In the European continent, the mechanical structure and operation of that technology confirmed in the eyes of contemporaries the neutrality and superiority of an authoritarian conception of order.[7] Political order, like the regularities and order produced by the clock, is superior when it has a maker. In the same vein, that the self-regulating clock was constructed by a maker served as a metaphor for God as the ultimate source of authority.

   The English, on the other hand, vigorously rejected the metaphor of the clock as a validation of authority. In England, where liberalism was experimented with in political life, the clock (and other automata technologies, as these were called) came to play a central role in reaffirming the liberal idea of a spontaneous, self-regulated order. Most notably to our discussion here, Adam Smith and David Hume found in the operation of self-regulated machines the material affirmation for the liberal operation of the market, which they could only have envisioned in the abstract (see Mayr 1986, chap. 10). For both Smith and Hume, it was liberty, rather than authority, that was affirmed by these machines.

"Upgrading" Habermas's Market Legitimization

As we can see, technology discourse is well linked with the history of political legitimation. In his essay, "Technology and Science as 'Ideology,'" Habermas (1970) lays out a history of market legitimation whereby a legitimation based on the principles of political economy à la neoclassical economics, that is, on the internal workings of the market, is replaced by another, external legitimation with the emergence of the welfare social democratic state and, more broadly, the emergence of Fordist society. It is from this point onward that political practice is measured in terms of the technical problems at hand rather than in substantive terms. The role of politics is reduced to finding the technical means to achieve goals (such as economic growth) that are in themselves understood to lie outside the realm of politics (Habermas 1970, 100–103). Technology is ideological to the extent that political issues are treated as technical issues: tensions and contradictions are overcome by delimiting the scope of the political, and, as a result, the instrumental rationality of technical language colonizes the sphere of politics.

Habermas writes at the zenith of Keynesianism, when the state is still very much involved in the administration of the capitalist economy. This administration of the economy is in fact the source of political legitimation. But this is politics "light," in Habermas's view; it is politics limited to technical administration: "Insofar as government action is directed toward the economic system's stability and growth, politics now takes on a peculiarly negative character. For it is oriented toward the elimination of dysfunctions and the avoidance of risks that threatens the system: not, in other words, toward the *realization of practical goals* but toward the *solution of technical problems*" (102–3). Habermas further outlines the circular and reinforcing relations between technical rationality and politics in the context of the welfare state. Incidentally, he also launches a critique on the postindustrial thesis and its underlying technologistic bent. With minimal modifications, this description might also apply to the operation of the digital discourse in contemporary political culture:

> It is true that social interests still determine the direction, functions and pace of technical progress. But these interests define the social system so much as a whole that they coincide with the interest of maintaining the system. *As such* the private form of capital utilization and a distribution mechanism for social rewards that guarantees the loyalty of the masses are removed from discussion. The quasi-autonomous progress of science and technology then appears as an independent variable on which the most important single system variable, namely economic growth, depends. Thus arises a perspective in which the development of the social system *seems* to be determined by

the logic of scientific-technical progress. The immanent law of this progress seems to produce objective exigencies, which must be obeyed by any politics oriented toward functional needs. But when this semblance has taken root effectively, then propaganda can refer to the role of technology and science in order to explain and legitimate why in modern societies the process of democratic decision-making about practical problems loses its function and 'must' be replaced by plebiscitary decisions about alternative sets of leaders of administrative personnel [i.e., technocracy] . . . it can also become a background ideology that penetrates into the consciousness of the depoliticized mass of the population, where it can take on legitimating power. It is a singular achievement of this ideology to detach society's self-understanding from the frame of reference of communicative action and from the concepts of symbolic interaction and replace it with a scientific model. Accordingly, the culturally defined self-understanding of a social life-world is replaced by the self-reification of men under categories of purposive-rational action and adaptive behavior. (105–6)

Science and technology are presented as the central axis of social development; this in turn requires the technical administration of system; finally, questions that should have been addressed politically by citizens (through communicative action) are subjected to purposive rational (i.e., technical) considerations to be resolved by system experts. The result of this postindustrial, or technocratic, condition is "the disappearance of the difference between purposive-rational action and interaction from the consciousness not only of the sciences of man, but of men themselves. The concealment of this difference proves the ideological power of the technocratic consciousness" (107). This is then an ideological phenomenon: "The ideological nucleus of this consciousness is the elimination of the distinction between the practical and the technical" (113)—that is, the elimination of the distinction between substantive (ends) and instrumental (means). Technology as ideology "severs the criteria for justifying the organization of social life from any normative regulation of interaction, thus depoliticizing them. It anchors them instead in functions of a putative system of purposive-rational action" (112). The depoliticizing effect of this is quite obvious: "The new politics of state interventionism," he says, "requires a depoliticization of the mass of the population. To the extent that practical questions are eliminated, the public realm also loses its political function" (103–4).

Now, with the shift to post-Fordism—that is, with the marketization of society and the disorganization of the economy—technology discourse in contemporary times no longer serves as the legitimation for political power to technically manage the capitalist economy. Instead, it serves as the legitimation for political power to *take a step back* from the capitalist economy. With the rise of neoliberalism as the economic dogma of contemporary society, and as the state withdraws from the economy, market legitimation has now returned to what Habermas identifies as the old model of market legitimation—classical political economy based on the internal workings of the market—but with a technologistic twist.

It is in this respect that the digital discourse is most crucial. It offers a renewed confidence in the market as a superior medium of economic and social life based on its improvements by technological means. The reason for the state to recede, and for the market to dominate is due to the materialization and perfection of the workings of the market by technological means. The legitimation of the market can rest entirely on technology. Contemporary technology discourse legitimizes not the intervention of the state in the economy but instead its withdrawal; not the external managing of the market, but the need of politics to let the market self-regulate. The goals have changed, but the underlying depoliticizing ramifications of technology discourse that Habermas was concerned with still persist.

It is in this context that the narratives of the digital discourse discussed in this chapter (as well as in the following chapters) are relevant. The ideas of spontaneous order, infinite prosperity, classless society, and market democratization—*woven into network technology*, integrated into the very medium where the social now takes place—reassert what for the good part of the twentieth century has been harshly criticized: the superiority of the market—open, frictionless, unhindered, and most importantly insulated from any political intervention—as a medium for social relations. In this respect, the digital discourse has the same ideological thrust as neoliberalism, as succinctly identified by Duggan (2003): "The most successful ruse of neoliberal dominance . . . is the definition of *economic* policy as primarily a matter of neutral, technical expertise. This expertise is then presented as separate from *politics* and *culture*, and not properly subject to specifically political accountability or cultural critique" (Duggan 2003, xiv). Previous critiques of capitalism (most notably Marx's) were aimed at the political economy underlying market legitimation, from Smith to Hayek. Through recourse to a technologistic framework, the digital discourse offers the rhetorical means by which neoliberal tenets are upheld and the critique of capitalism is bypassed.

Markets are not the only systems that are transformed by network technology; the whole system of production changes as well. The next two chapters present the digital discourse on these transformations. Chapter 4 focuses on the ramifications of network technology for the traditional (or industrial) world of work, mostly within the confines of the workplace and the company. Chapter 5 focuses on new forms of work and production brought about by network technology and taking place outside of these traditional sites.

CHAPTER 4

NETWORK WORK

IN CRITICAL SOCIAL THEORY, TECHNOLOGY OCCUPIES A CENTRAL ROLE in the organization and distribution of power. Foucault, for example, sees the capilarization and disciplinarization of power in modernity as tied to science and technology. The shift from overt authoritative control to implicit disciplinary control entails the emergence of disciplining techniques. Subjectification and self-disciplining through the construction of a panoptic space is a prime example of such technique (Foucault 1995). Technologies, even in the narrowest sense of tools or machines, are a particular case of Foucault's indivisible knowledge/power nexus; they are a phenotype of both knowledge and power and cannot be analyzed solely in terms of one or the other (Foucault 1994). More recently, Bruno Latour upheld the indivisibility of knowledge—in the form of technology—from social relations, stating that "technology is society made durable" (Latour 1991), and Castells concurs, saying "technology *is* society, and society cannot be understood and represented without its technological tools" (Castells 1996, 5).

A somewhat similar position regarding the nexus of technology and power can also be found in the Marxist tradition. Here, technology is seen as part and parcel of a class struggle. For Marx, the mode of production is a product of a dialectical play between relations of production and forces of production. In this formulation, technology, as a force of production, constitutes and is constituted by social relations (see Figure 4.1).

Hence, research along Marxist lines sought to unravel this nexus and uncover the component of power within technology. It sought to show how technology was not simply a force of production, that is, a rational, universal tool waiting to be deployed for the service of humans, as the Weberian framework of rationalization suggests (Weber 1978, 67; Hård 1994, 16), but rather is submerged in social relations and is a tool of class struggle. Most research on technology in the Marxist tradition focused on technologies of production, assuming that the

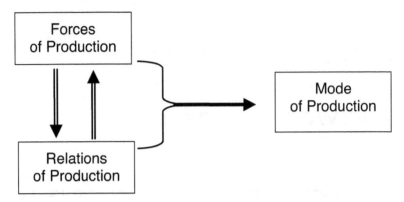

Figure 4.1.   Marx's formulation of the mode of production as constituted by the dialectical relationship between productive forces and productive relations

reproduction of capitalism took place within the confines of the work process itself. Groundbreaking studies, such as Harry Braverman's and David Noble's, have shown that major technological developments were introduced into the labor process as part of a class struggle. Braverman's *Labor and Monopoly Capital* (1974) analyzed Taylorism and the fragmentation of the labor process in terms of the deskilling of workers, that is, the dwindling down of the most precious resource that labor has *vis-à-vis* capital, its skills. Noble's *Forces of Production* (1984) further showed that the technologies of choice in the workplace were those that transferred more control over the production process from workers to managers. These choices reflect not simply the dictates of technical rationality, but stem from a compulsion, rooted in capitalism, for total control over the labor process (Feenberg 1991, 35). The ramifications of deskilling and diminishing control of labor over the work process brought about by technology was well understood by eighteenth-century Luddites, either tacitly or explicitly (Robins and Webster 1985). In short, and as numerous other studies along this line have shown, technologies play a central role in the power struggle between workers and capital, with the general trend being the weakening of the former vis-à-vis the latter (Aronowitz and DiFazio 1994; Aronowitz 2001; Huws 2003; Greenbaum 1995; Dyer-Witheford 1999; Noble 1995).

In this formulation technologies of production are central tools in the reproduction of social relations, and technology in the workplace plays a central role in the reproduction of a capitalist class structure throughout the industrial era. But how, according to the digital discourse, does this nexus of technology, work, and class change with the introduction of network technology into the equation? What does the new nexus look like? This chapter examines the digital discourse at the intersection of network technology and the world of work. It asks what, according to the digital discourse, happens to the workplace, and the world of work in general, upon its integration into network technology. This chapter explores these questions through three narratives. The first narrative concerns the blurred boundaries between work and leisure, work space and life space, and work time and private time. The second narrative concerns the blurred boundaries between capitalists and workers, bosses and employees, that is, those who own the means of production and those who do not. These two narratives, I suggest, downplay the significance of power in the new informational mode of production and give prominence to network technology as the sole coordinate of social power. This conceptual move is further explored in the third theme, which highlights the prominence of professionalism and meritocracy and the centrality of network technology in their construction. Network technology becomes for the informational worker not only an instrument and a vehicle to define his class position but also the prime source of the self, a central axis to define the worker's social position.

The chapter concludes by offering a theoretical framework to understand these new narratives. Network work, according to the digital discourse, creates a new constellation of technology, work, and power. The workplace become more flattened and dehierarchized, work becomes more decentralized and eroticized, and workers become more liberated. In contrast to industrial technology and the organization of work that sprouted from it—which were oppressive, heavy handed, and authoritarian, rendering workers cogs in a big machine—network technology and network work are more liberating, allowing more personal expression and freedom. These transformations in the workplace are presented by the digital discourse as undermining the industrial notion of class. These new narratives constitute a new spirit of networks that locates network work as a response to the humanist critique of industrial technology and industrial capitalism and that, at the same time, downplays a response to the social critique. In other words, the discourse on network work legitimates a new trade-off between alienation and exploitation: as it promises to mitigate alienation, it conditions this mitigation by the exacerbation of exploitation.

The nexus of network technology, work, and class in the digital discourse can be illustrated through the analysis of a new persona, or a new subjectivity, that permeates the digital discourse on work: the digerati. The term "digerati" is a compound of "digital" and "literati" and denotes, according to Wikipedia,[1] "the elite of the computer industry and online communities." The digerati are the carriers of the "spirit of informationalism," in Castells' terms (1996, 195), or the spirit of networks, in mine. The digerati—mostly men (Stewart Millar 1996)—work in and around information technology, usually at the forefront of the development, financing, and deployment of that industry. The digerati, and the social group to which they belong, have been variably referred to as "informational workers" (Castells 1996), "symbolic analysts" (Reich 1991), the "virtual intelligentsia" (Lovink 2002), the "virtual class" (Barbrook and Cameron 1996), and the "creative class" (Florida 2003). The term "digerati" is frequently used in *Wired*.[2] Indeed, *Wired* addresses chiefly the digerati and plays a central role in the articulation and construction of the identity of this new social category. The digerati are defined first and foremost by their work, specifically their engagement with network technology. Their presentation in the magazine is therefore an appropriate locus to explore the new nexus of network technology and work.

One of the most elaborate and crystallized portrayals of the digerati in *Wired* can be found in "Microserfs" (Coupland 1994), a fictional story about seven Microsoft employees told in the first person by one of them. This literary piece (written by novelist Douglas Coupland, a leading voice in the digital culture, and later published as a book with the same title in 1995) encapsulates the "essence" of the digerati as a new social category, specifically, as a new type of worker. Let us follow the subjective experiences of the informational worker as represented in "Microserfs," as well as a few other articles in the magazine.

THE BLURRED BOUNDARIES BETWEEN WORK AND LEISURE

According to the digital discourse, network technology facilitates the blurring of the boundaries between work and leisure, between productive time and free time, between office space and home space. This means that work is ever present and can be performed everywhere, but it also entails the infusion of creativity, joy, and personal expression into work and, in turn, the dealienation of work. The digerati is a new subjectivity that embodies a profound transformation in the notion of work from the Fordist to the post-Fordist workplace. It is no longer the image of the worker going from his *home* space to his *work* space in order to *labor*, only to retreat afterward to his *leisure* activities. The distinction between work and leisure is thus blurred.

This blurring is a double-edged sword. On the one hand, this blurring commonly entails the colonization of leisure by work to the extent that work seems to define and structure the digerati's personal lives.

One indication of that is simply the amount of time that the workers in "Microserfs" spend at work. While one of their colleagues is reported to be locked in his room at 2:30 a.m., trying to solve a problem with the code he is writing, the other Microserfs are heading to the supermarket, which is "completely empty save for us and a few other Microsoft people" (Coupland 1994). Elsewhere in the story, the protagonist wonders whether a co-worker—whose eyes are "all red and sore"—ever sleeps. But the penetration of work into leisure time is not just a quantitative matter: private life is overwhelmed by work life. As the protagonist of "Microserfs" grudgingly admits, "Microsoft is not conductive to relationships." He tries, he acknowledges, to get other things going, "but then work takes over my life," leaving him with "work, sleep, work, sleep, work, sleep" (Coupland 1994). This description is reminiscent of Marx's lament about the alienating effects of (industrial) labor, where the worker "feels himself to be freely active only in his animal functions—eating, drinking, procreating, or at most also on his dwelling and in personal adornment—while in his human functions he is reduced to an animal. The animal becomes human and the human becomes animal" (Marx 1995, 99).

But for the most part, the blurring between on duty and off duty is seen as positive progress by the digerati and as a trend that allows the injection of creativity, joy, and personal expression into work, or the "erotization" of the workplace in Marcuse's (1974) terms (see also Agger 2004). The blurring of work and leisure can be seen, for example, in the production of new workspaces that openly incorporate leisure ethics and culture. This is evident at a symbolic level: work no longer takes place in a factory but on a "campus" (Coupland 1994). Work is a place where "research and development," "intellectual work," "creative work," and the "analysis of symbols"—to name just few of the neologisms associated with network work—substitute labor. Work in a campus is not so much about production as it is about creativity and contemplation.

This blurring is also evident on a spatial level. An illustration in a special issue of *Wired*, guest edited by renowned architect Rem Koolhaas, shows the network workspace to be explicitly designed to blur the distinction between the rationalized, productionist, public, and disciplined world of work and the eroticized, irrational, consumerist, and private world of leisure.[3] This space combines elements of work (or social *reproduction*), such as computer stations and small cubicles, with elements of a more social and eroticized space (pertaining to social *communication*), like a coffeehouse and a living room; as the text accompanying the image explains, "the lobby looks like a living room, a cluster of workstations like a café" (Koolhaas 2003). The title of the article—"Where Do You Want to Work Today?"—implies a more lenient, flexible approach to work. It is not only that workers can decide where it is they are going to work: in front of a computer, at the coffee bar, or on the couch in the living room; more important, perhaps, is that work itself is reconstrued in terms of leisure, involving—in this case—personal, consumer-like choices (where do *you* want to work? the "cubicle," the "café," or the "living room"?—it's your choice). At the same time, these sites of leisure are also reconstituted as spaces of work.

The blurring of the time and space of work and that of leisure, with the liberating undertones of a workspace that feels like home or a coffee shop, is reiterated in two advertisements for mobile technology. Their analysis also allows a glimpse into the unvoiced and underplayed hegemonic meanings of such blurring. In both, mobility—afforded by network technology—implies more opportunities and more freedom by liberating individuals from the constraints of the traditional workspace. But what is the nature of this freedom? What *kind* of emancipation does mobile technology promise to deliver?

The first advertisement, for Verizon Wireless (2004), depicts a woman dressed in a business suit seated on a platform of a train station and working on her laptop. The advertisement's copy reads, "Even if the train's late, your work gets out on schedule." The digerati, the new informational workers, process information rather than matter and are therefore able to break free from the industrial anchors of space and time, from the shop floor of production. But in this freedom, afforded by network technology, lies also its dialectics. Inasmuch as the digerati are liberated from the particular space of work, they also potentially become workers at all times. The *mies-en-scène* of the advertisement could just as well be the woman's bedroom before she goes to sleep or a hotel room on her vacation. The digerati, who no longer subscribe to the old dichotomy of work and leisure, work time and playtime, find themselves at work, or in a *capacity to work*, at potentially all times and all spaces. It is not only the café that penetrates the workspace; it is also work that penetrates the café.

The opportunity and freedom afforded by mobile technology links leisure to work in a manner that makes their demarcation more flexible and makes it easier for companies to colonize the time of their employees by making the technologies of production (such as mobile phones, laptop computers, and above all, Internet connectivity) available to workers at all times. As the text openly says, "You can always count on Verizon Wireless . . . So keep *productivity on schedule* across the U.S." (Verizon Wireless 2004; emphasis mine). The blurring boundaries between work and leisure results in the expansion of productive time and space; mobility turns out to be a new force of production, and freedom turns out, in this case, to be the freedom to work during one's leisure time.

Hence, the blurring of the distinction between work time and leisure time, and workspace and leisure space means that more and more time—at home, on the way to and from work, and on vacation—can be consumed by work. At the same time, this also means that work itself is no longer construed as the negation of leisure and freedom. This narrative comes out of a two-page advertisement for T-Mobile (2004). Here, too, the subtext seems to reveal a dialectic that the surface visual and textual presentation attempts to plaster. The left-hand page of the advertisement shows a gloomy, florescent-lit office with two rows of identical cubicles and office chairs along opposite sides of the office space. The opposite page uses an identical visual composition, only this time the chairs are substituted by two rows of trees, with a group of bikers riding on a scenic road amid a beautiful green landscape. The text reads, "Out. Brought to you by T-Mobile."

The advertisement suggests that, thanks to mobile technology, the digerati are not anchored to the locality of the workplace and are therefore liberated from the drudgery of a bureaucratic nightmare: free from office hours and office space, from supervisors and routine, from the process of mass production, they are cogs liberated from the machine. These grievances are, of course, at the heart of the humanist critique regarding the alienating ramifications of industrial work to which the digerati's spirit of network responds. But further contemplation reveals once again the dialectical meaning of this liberation: with network technology, it is also *work* itself that is now liberated from space, rendering all spaces workspaces. The blurring of the differences between the enclosed office, on the one hand, and "out," on the other hand, also entails the surrender of personal time to corporate time—the surrender of "out" to the rationale of "in." So it is true that one can work while enjoying the outdoors and keep a tab on business while going on vacation. But, at the same time, the very essence of what it means to be "outdoors" and "on vacation" is transformed completely.

The freedom brought about by mobility entails not only the intermingling of work time and playtime but also the freedom of work time to colonize (symbolically and practically) playtime. The advertisement initially seems to suggest that liberty from workspace also implies liberty from work itself; laying the two contrasting visual images next to each other implies that mobile technology transfers us from the world of work (represented in the left image) to the world of leisure (represented in the right image). Again, the dialectical nature of this transfer does not evade the story that the advertisement is telling: the whole point of the plot in the advertisement is that while one is away from work, work does not need to be away from one or that the opposite of "working indoors" is "working outdoors" (rather than not working at all). In that formulation, even though the two worlds—that of work and that of leisure, the indoor and the outdoor—are presented as a mirror image in the ad, they become *functionally* the same space: that of production. This narrative is reiterated in a myriad of other advertisements, all of which extol network technology as the centerpiece of a more mobile, flexible worker, emancipated from the constraints of time, space, and structure emblematic of the oppressive industrial, bureaucratic organization.

THE BLURRED BOUNDARIES
BETWEEN WORKER AND CAPITALIST

The blurring between work time and workspace, on the one hand, and leisure time and leisure space, on the other hand, is coupled with another narrative regarding the revolutionary nature of network work. Network work also entails the reconstitution of workers as individual nodes within the network, a mirror image of workers in the Fordist organization. The Fordist workers were conceived as part of a system: functionally, they were cogs in the machine; structurally, they were members of a social class. Both of these imply *hierarchy* and *differentiation* between functional levels of the organization, between management and workers, between tasks, and, ultimately, between capital and labor. The digerati embody a critique and an alternative to hierarchy and differentiation at work and in society. In the network workplace, hierarchy and differentiation are replaced by a network structure where information flows between individual nodes, with each node structurally homologous to its peer and capable of multitasking and entrepreneurship. Workers are therefore conceived in more individualized and privatized terms rather than in terms of members of a class.

The new informational worker, in fact, resents the modernist, industrial-age distinction between capitalists and proletarians, a distinction formulated at the height of industrial capitalism and deemed no longer applicable. When information technology becomes the center of the economy, says Peter Drucker, information and knowledge rather than capital, and informational workers rather than capitalists, take center stage. Since knowledge substitutes capital as "the new basis of wealth," since information displaces "the traditional factors of production—land, labor, and capital—[which] are becoming restraints rather than driving forces," contemporary society is better understood as "post-capitalist" (Schwartz 1993). For Drucker, the emergence of "knowledge, not capital, [as] the new basis of wealth" also signifies the downfall of the blue-collar working class, which "dominate[d] every single developed society"

(Schwartz 1993). The ushering in of new forces of production, in the form of information, knowledge, and information technology, transforms the very basis of social class. It is noteworthy that Druker's theorization follows in the footsteps of Daniel Bell's thesis regarding *The Coming of Post-Industrial Society* (1999).

What substitutes the blue-collar working class in the digital discourse on network work is a new persona, no longer tied to the traditional industrialist conceptions of a proletariat or working class. At the phenomenological level, work relations have been dehierarchized to such an extent that the differentiation between mangers and workers, employers and employees, is blurred. In the fictional story "Microserfs," the narrator refers to his superior at work as "my 'boss'" (Coupland 1994); the quotation marks around "boss" indicate the term is actually a misnomer, a relic of power relations that no longer exist. Power structures, he insists, are now completely flattened and dehierarchized: "The person with the most information pertinent to any decision is the one who makes that decision" (Coupland 1994). Upholding the meritocratic ethos of the digerati (to which I return later in this chapter), power in the network workplace is presented in the digital discourse as a corollary of knowledge and expertise, not of capital.

Indeed, money and even capital (i.e., money diverted to further capital accumulation, rather than for consumption) seems an ill-fitting variable to distinguish employers from employees according to the digital discourse. The protagonist of "Microserfs," while nominally a worker, constantly checks his "WinQuotes" on his computer screen to follow Microsoft's NASDAQ price. "It was Saturday," he recounts, "and there was never any change, but I kept forgetting. Habit. Maybe the Tokyo or Hong Kong exchanges might cause a fluctuation?" Workers are also stakeholders in the company; in the case of "Microserfs," they are actually shareholders. "Most staffers," he continues, "peek at WinQuote a few times a day. I mean, if you have 10,000 shares . . . and the stock goes up a buck, you've just made ten grand!" (Coupland 1994). Such positioning of the digerati—not quite capitalists, not quite proletarians—undermines previous notions of class. Should they be better understood as workers or as owners? Or is the very distinction nothing more than an industrial-era relic that has been blurred beyond distinction and does not hold for the digerati? Let us look more closely at the narratives regarding the two sides of the coin: the digerati entrepreneurs and the digerati workers.

THE DIGERATI ENTREPRENEUR

On one side of the informational coin are the digerati entrepreneurs. Just as the working digerati try to distance themselves as far away as possible from industrial workers, so does the entrepreneur digerati construct their identities as a reverse image of the persona of the industrial capitalist. The industrial capitalist is characterized by his passion to accumulate capital; digerati entrepreneurs, in contrast, are characterized by their passion for network technology. The former sees technology as a means of production, that is, as a means to another end. The latter see the development of technology as an end unto itself. Even the one criterion that might have put the old industrial capitalist and the new digerati entrepreneur in the same social category—their wealth—is seen in the digital discourse as qualitatively different. Referring to the founders of Microsoft, the narrator in "Microserfs" argues that "Microsoft's millionaires are the first generation of North American nerd wealth" (Coupland 1994), distinguishing their "nerd" wealth from that of previous, industrial-age moguls. Not unlike the way "old money" (based on inheritance) was distinguished at the beginning of the industrial era from the "*nouveau riches*" (based on entrepreneurship).

According to the digital discourse, digerati entrepreneurs are different from industrial-age capitalists because for the digerati, business does not all revolve around profit but on contributing to technological progress; their success is measured not only in terms of material gains but also in terms of their technological skill, or of how "techie," "geek," and "nerd" they are. The August 2004 cover of *Wired*, for example, is adorned by a teaser that reads "The Top 100 Geeks"; two out of the four geeks mentioned on the cover are iconic digerati entrepreneurs: Bill Gates and Steve Jobs (Cover 2004). It is precisely the dehierarchized conception of network work that makes it plausible to describe two of the most powerful and rich people on the planet as underdogs, an idea I will return to later in this chapter.

In the digital discourse, reverence is directed toward the entrepreneur-*cum*-technological nerd. The real incentive for the new informational entrepreneur to engage in business, according to the digital discourse, is not making more money but making a real social contribution: discovering new opportunities for network technology and opening up new frontiers for the network society. In more theoretical terms, technology, as a means of production, takes center stage, while ownership over these means, the component of power in the technological nexus—that which distinguishes classes—recedes to the background, leaving only vague traces.

The digital discourse's portrayal of the new digerati entrepreneurs—industry moguls like Bill Gates, Steve Jobs, or dot-com millionaires of a more modest caliber—usually follows this narrative. Here is one example from *Wired*. Josh Harris is a thirty-nine-year-old dot-com millionaire who now invests his fortune in a new revolutionary media project that is unlikely, according to his own estimates, to yield much, if any, profit. Harris assesses that he "could have gone into anything after Jupiter [where he made his fortune]. I could have been an *industrial titan* . . . But I didn't want to be like *those guys*" (Platt 2000; emphasis mine). What are "those guys" like? According to Harris, "Most wealthy people just waste their time making more and more money." In contrast, he says about himself, "I'm one of the few who actually know how to spend it." The author of the article concurs: "Making money . . . is not Harris' primary goal these days. He's too busy spending it" on unprofitable, but technologically revolutionary, projects (Platt 2000). Harris, like Weber before him, laments the iron cage of capitalism with its internal drive for the endless accumulation of capital. However, he not only criticizes it but also posits an alternative to this spirit of capitalism, one where the entrepreneur is motivated by the endless progress of technology.

With such new spirit, it is hardly a surprise that the new informational entrepreneur—committed to creativity and discovery—is at severe odds with the strict, profit-driven, industrial-age corporation. Hence, the digital entrepreneur is also a rebellious character, a contrarian. Describing his experiences while running a small research and development lab, Harris recalls the clash of his new digerati spirit and the old corporate mentality: "At first I made it [the lab] a pure art house . . . I mean, why not? How often do you get the chance to do that kind of thing? The price I paid was that I had to deal with the corporate people at Prodigy . . . Executives at [Prodigy] . . . understood that they were dealing with a *loose cannon* who had little respect for their *corporate culture*" (Platt 2000; emphasis mine). The persona of the digerati represents a break with previous corporate culture. In the digital discourse on network work, the digerati are constructed as reverse images of the industrial man, indeed as an embodiment of a critique and a repudiation of him.

It is also worth pointing out precisely the *depth* at which the new spirit breaks with the old and where the two are actually agreeable. If Harris was so subversive to Prodigy's "corporate culture," one might wonder why they nevertheless put up with him. According to Harris, "They didn't want to fuck with me because I was accounting for 25 percent of the total hourly traffic on their service" (Platt 2000). Hence, the divide between new and old capitalists that Harris was referring to seems to be located at a cultural and stylistic, rather than a structural, level. Notwithstanding the presence of the "old-fashioned" bottom line in Harris's narrative of the new digerati entrepreneur, the façade is casual and subversive, decidedly antiestablishment, an establishment of which he is (perhaps grudgingly) a part.

This opposition between the old corporate culture and the new digerati culture is reiterated in the article when the author describes Harris's current workspace as the opposite of what one might expect a corporation to look like: "an open-plan office that looks like it was furnished by hyperactive children turned loose in a thrift store. Miscellaneous desks and tables are crammed together," and so forth (Platt 2000). The digerati entrepreneur embodies this tension between, on the one hand, a disorganized, creative, impulsive attitude ("Harris once again became bored with the grind of managing a company . . . [and] surrendered his position as CEO") and, on the other hand, a business-oriented mindset committed to capitalism after all. This tension is, according to Harris's assistant, part of who Harris is, an emblem of the new entrepreneurial digerati: "Josh is totally disorganized," says his assistant, "but he *chooses* to be disorganized, to free his *creative mind*. Underneath, there is a *solid core of business sense*" (Platt 2000; emphasis mine). Strong personal expression, disorganization, a mischievous attitude, and creativity—all that the old corporate culture asked to check in at the entrance to the office and that were more appropriate to entertain during leisure time—are not just tolerated in the digerati culture but embraced and unleashed as new forces of production.

The persona of the digerati entrepreneur blurs the distinction between the capitalist and the worker in another way: by mimicking the worker and standing shoulder to shoulder with him. The digerati entrepreneurs no longer see themselves as a representative of the system; quite the contrary, the digerati entrepreneurs grasp themselves as antagonist to the system. They are antistatist, antiestablishment, antibureaucracy, and anti-corporation, just as the informational worker is. They are *ontologically* the little guys, the underdogs, the counterculturists, and the tramps, mobilizing their individual faculties to fight the oppressiveness of the powers that be.

This duality within the digerati entrepreneurs—capitalists-businessmen-bosses, on the one hand, counterculturalists-radicals-hackers, on the other—surfaces in an interview with another digerati entrepreneur—Tom Jennings. His central characteristic is exclaimed in the title of his *Wired* portrait, "The Anarchist" (Borsook 1996). Jennings is described as having a shrewd business sense, technological savvy and creativity, and radical views about both business and technology. The article portrays the business side and the technology side of the high-tech industry as if they were utterly conflicting and contradictory to each other. In this dichotomous framework—business and capital versus creativity and

technology—the digerati are associated with the latter and virtually alien-
ated from the former. Right from the outset, "doing business" in general
is portrayed in the article as fundamentally different from "doing digital
technology-related business": "It seems these days," the author laments,
"that to succeed as an anarchist"—Jennings's central characteristic—"you
have to succeed as a businessman." Jennings explains that in the founding
days of the high-tech industry it was not as business-oriented as it is now-
adays: "With early microcomputers, it was more exciting—everyone was
a weirdo. Then it got boring, all suits. The suits in the Internet business
now are taking chances, and even the worst of them have some interest in
what's going on. But in a few years . . . they'll be terrible" (Borsook 1996).
Like Harris, Jennings—notwithstanding his riches, his involvement with
business, and his position of power—self-identifies in contrast with all
of these, indeed in defiance of them. The author sums up his contrarian
demeanor this way: Jennings, he writes, "never relinquished his lifelong
irreverence toward TPTB—The Powers That Be. He's an anarchist, pure
and simple—which means he professes to no ideological purity of any
kind. He espouses no doctrine . . . about left or right, or about the marketplace" (Borsook 1996).

The flattened hierarchy of the new network workplace and its embodi-
Jennings, like Harris and Gates, is portrayed as a heroic figure, an ideal
type for the digital age. They are all "great men," in the words of Boltanski
and Chiapello (2005, 112): they define, embody, and champion the qualities of those who are at the forefront of capitalism and are the ones up
to whom all other actors look. Jennings's prowess and greatness is traced
back to technology rather than capital, to machines rather than social
relations; technology, and engagement with technology, becomes an axis
of social power, and structural class power is neutralized and recedes to
the background. "Probably without ever meaning to," the author writes,
"[Jennings] always lands on the cutting edge of what's considered cool
and now and happening . . . because he's so damn technosmart and
blessed with such sound instincts" (Borsook 1996). The new company
he now runs focuses, according to the author, on "technological savvy,"
and Jennings adds, "We terrify people because we're so technically astute"
(Borsook 1996). Being "technosmart," "technological savvy," and "technically astute"—all highlight the direct engagement of the digerati with
network technology—concurrently imply the underplaying of power and
class that is entailed by business ownership and business management; in
the digital discourse, "techiness" *topples* class.

The flattened hierarchy of the new network workplace and its embodiment in the persona of the digerati entrepreneur are succinctly sum-
marized in two images that accompany *Wired*'s profile of Andy Grove

(Heilemann 2001). We should perhaps recall here the stature of Andy Grove in the digital industry. Andy Grove is the legendary founder and longtime president, CEO, and chairman of the board of Intel, the largest processor maker in the world with an annual revenue of 38.8 billion in 2005, for which it was rated forty-ninth in the *Fortune 500*. In 1997, Grove was *Time* magazine's "Man of the Year." He is certainly a "great man" in the new capitalism. Notwithstanding Grove's heroic and superstar position in the new capitalism, he is portrayed in the digital discourse as an Everyman, a little guy; even Andy Grove is a node in the flattened network workplace.

The first image features Grove, at a conference at Stanford Business School, behaving in a manner that would clearly be atypical for an industrial-era tycoon but that typifies the new digerati entrepreneur: he is casually sitting on the floor, suit jacket off, relaxed and listening. The second image is even more explicit, since it depicts not just a fleeting moment at a conference but rather a more permanent reality: Grove's workspace. The photo carefully captures a wide-angle perspective of Intel's workspace: a multiplicity of cubicles in a shared work floor. Andy Grove, this "great man," is visually rendered as small, photographed from above, and as part of the hive of workers in the network work. The title of the photo tells that this is "Grove's 8' by 9' cubicle at Intel," implying that Grove is treated like any other worker (Heilemann 2001). The flattened, dehierarchized network workplace is perceived as an emblem of a new social structure that is better defined in terms of the network than in terms of class. (Little surprise, then, that in this workplace without differentiation, a quick entrepreneur is marketing a new line of clothing to fit the new persona of the digerati, described as "fashion for the no-collar workforce" [Glickman 1999]).

But while the structural tension (or even the personal animosity) between employers and employees is presented as a thing of the past, there are still reminders in the text of the relational positions of these two categories within the company and within society at-large. At these moments, the elasticity of the narrative of a new network-*cum*-classless society seems to fail, and the persistence of class in the information society and the high-tech company surfaces. One indication of that is the harsh language used by the protagonist of "Microserfs" to refer to work. He tells, for example, of how the group "*slaved* until 1:00 a.m." Of course, there is the very use of the term "Micro*serfs*" (Coupland 1994; emphasis mine), which evokes both the highly advanced Microsoft Corporation and the highly regressed system of medieval feudalism.

With this subtle choice of wording, the author seems to point out the complexity of what is presented throughout much of the digital discourse as an unproblematic blurring of the dichotomy between workers and employers. While indeed far removed from the terminology of industrial working relations, the reference to slavery and serfdom also connotes feudal relations with *its* blurring—indeed, annihilation—of the distinction between work and leisure, between private and public life. Perhaps, these terms suggest, these new working arrangements and work ethics are taking us back, not forward. But such moments of critical reflection are relatively rare–slips of the tongue. For the most part, the digital discourse narrates the transformations of network work quite uncritically.

THE DIGERATI WORKER

Just as capitalists assume the role of workers, workers, too, perceive themselves more like capitalists and entrepreneurs. As we have seen in Chapter 3, market flux and instability render work more flexible and dynamic. Work for the digerati no longer follows the industrial-age model of a stable career path, which entails a lasting time commitment between workers and companies and opportunities for advancement within the firm. According to Kevin Kelly, the decreasing ratio of companies to workers in the market reveals a trend in which more and more individuals become involved with more than one workplace, or job: "So many workers double- or triple-dip, moonlighting, part-timing, consulting, or running a startup in their garage." This signifies a qualitative transformation in the world of work from the old employment model to what Kelly calls "polyemployment" (Kelly 1999). Work is redefined in terms that are more individualized and privatized, virtually eliminating the distinction between worker and company, employment and entrepreneurship, and—as a corollary—the distinction between classes. After all, workers can freelance as consultants and, on top of that, be entrepreneurs running their own start-up.

These new conditions inevitably make the experience of a career more liquid, precarious, insecure, and fragmented. The fictional story "Microserfs" reveals the digerati's anxieties that stem from unpredictable career prospects in the new economy. While being comparatively well paid (with a compensation packet that often includes stocks and options) and enjoying a high social status, the digerati are threatened by no less (perhaps even more) job insecurity than wage laborers in other sectors. Thus, at some moment in the story, the protagonist of "Microserfs" laments the ageism prevalent in the digital sector. After a visit to the Nintendo headquarters, located right across from Microsoft's, he comments, "It's like the year 1311, where everyone over 35 is dead or maimed." Elsewhere he makes the same observation about Microsoft, saying that there are "maybe twelve fortysomethings on the Campus" (Coupland 1994). Another article describes one Matthew Clark, who left Microsoft, following "a severe case of corporate burnout" (Glickman 1999), when he was only twenty-seven years old.

Thus, anxiety about old age and about the future begins almost at the outset of the digerati's careers. Referring to a group of new employees at Microsoft, the protagonist of "Microserfs" says,

> [We] were wondering last week what's going to happen when this new crop of workers reaches its inevitable Seven Year Programmer's Burnout. At the end of it they won't have two million dollars to move to Hilo and start up a bait shop with, the way the Microsoft old-timers did. Not everyone can move into management. Discarded. Face it: You're always just a breath away from a job in telemarketing. Everybody I know at the company has an estimated time departure and they're all within five years. It must have been so weird—living the way my Dad did—thinking your company was going to care for you forever. (Coupland 1994)

Under such circumstances, the digerati worker is required to assume the mindset of an entrepreneur. If the undertone in "Microserfs" is one of lament, this critical edge is completely silenced in another piece, which more forcefully asserts the emergence of the worker-*cum*-entrepreneur and the concurrent demise of the distinction between workers and capitalists. Here, Tom Campbell, a former politician and the current dean of the Haas School of Business at the University of California–Berkeley, reflects on the golden age of Silicon Valley:

> The paradigm was creative destruction, or cannibalistic capitalism. To survive here, everyone needed to be mentally prepared to jump ship at a moment's notice. *The basic building block of the economy was the entrepreneur, and every individual needed to think more like an entrepreneur and less like an employee*... Jazzed by this new ethic, laid-back Generation X turned into the *self-determinist Generation Equity*. Average job tenure in 1999 was 15 months. We assumed that Moore's law applied to far more than chips. It applied to *everything*—bandwidth, user bases, even the Nasdaq doubled every 18 months... Silicon Valley was an argument in the form of a place. It argued for a new way to live, a new relationship between owners and employees, a new bond between work and play. (Bronson 2003; emphasis mine except "everything," emphasis in the original)[4]

For Campbell, the new dynamics of the high-tech sector have made workers into entrepreneurs, work into play, and wage laborers into shareholders, effectively rendering these distinctions obsolete. According to

Wired, network work renders each worker into a more individualized, independent node, who does not have a "job" in the industrial-era sense, but rather takes part in ad hoc projects. The spirit of networks is also the spirit of workers as entrepreneurs.

It is noteworthy that by 2003, Campbell suggests this dynamic was overturned, and more old-fashioned, industrial-age relations were reinstated: "The culture of shifting alliances and temporary agreements is out; permanence and settling down is in" (Bronson 2003). From Campbell's words it is clear that this entails a return to the idea of the "worker," a persona admittedly less sexy than that of the digerati and one that reestablishes its relational position in the mode of production.

TECHNOLOGY AS AN AXIS OF PROFESSIONAL IDENTITY

As we have seen thus far in this chapter, in the digital discourse on network work, the digerati are no longer industrial-type workers whose hierarchical positions are determined by inferior power relations *vis-à-vis* the capitalist and the capitalists, in turn, are also no longer the profit-driven industry titans commanding power and respect simply for their capital. These hierarchical and structural positions are replaced by flattened and networked relations facilitated by network technology. Network technology becomes the predominant axis in the tripartite nexus of work, technology, and class. As Daniel Bell puts it, power in postindustrial society is organized around a technological axis. This point is made even clearer in Bell's new forward for the 1999 edition of *The Coming of Post-Industrial Society*, titled "The Axial Age of Technology" (Bell 1999). The digerati's subjective position in the workplace and the economy-at-large does not so much stem from their relational position in the mode of production. Instead, their identity and social position are now determined solely by their location on a technological scale. The axis of their identities is now constructed around the dominant force of production, network technology; their subjectivity is mediated through technology and through their engagement with it.

In the digital discourse network technology becomes the central axis of class location and class analysis. Class location is determined by how high or low a person is on a "techie" scale. In other words, dehierarchization in the workplace, and, by extension, between capitalists and workers, does not annul the making of hierarchical distinctions altogether, but these are now based on the level of individual engagement and expertise with network technology, that is, they are based on meritocracy, or better yet "technocracy," or technological meritocracy. Thus, the digerati identify themselves as a distinct—indeed as an elite—social group. A reader of

Wired refers to readers of the magazine collectively as "techno-intelligentsia" (Olcese 2004).

This technology scale helps separate the in-group from the out-group, but it also stratifies the network workplace from within. Within the in-group of techno-intelligentsia, status is merit based; it is achieved through skillfulness and knowledge. But not just in any field; as "Microserfs" (Coupland 1994) reveals, reverence is held for skills and knowledge that go to the heart of network technology, such as programming. Referring to his superior, named Shaw, the narrator of the story says, "One grudgingly has to respect someone who's fortysomething and still in computers—there's a *core techiness there that must be respected* . . . My only problem with Shaw is that he became a manager and stopped coding. Being a manager is all hand-holding and paperwork—not creative at all. *Respect is based on how much of a techie you are and how much coding you do*" (Coupland 1994; emphasis mine). Hence, creativity—in this case, the ability to write codes well—becomes a virtue that "great men" are expected to possess. It is neither capital that commends respect, nor power, nor going up the ladder of management, nor even being a good businessman. It is not even dexterity as such that commends respect. It is only those skills that engage directly with network technology that serve as criteria for social stratification.

In that respect, it is interesting to note the figure of Bill Gates. In "Microserfs," it is made clear that Gates is revered (at least by Microsoft workers) not for his money or power but for his high level of techiness. The protagonist of "Microserfs" talks about Gates in almost transcendental terms: "The presence of Bill Gates floats about the Campus, semi-visible, at all times . . . Bill is a moral force, a spectral force, a force that shapes, a force that molds." The aura of Gates seems to radiate to more than just technological realms; among the community of Microsoft workers he is a cultural icon as well: "I bet if Bill drove a Shriner's go-cart to work, everybody else would, too." Elsewhere in the story, after having the rare opportunity of meeting with Bill Gates, one of the protagonist's co-workers reports, "People forget that he is medically, biologically, a genius. Not one um or ah from his mouth all lunch; no wasted brain energy. Truly an inspiration for us all" (Coupland 1994). According to "Microserfs," then, Gates is such a prominent figure not because of his social power but because of his techiness. Other attributes of Gates—his being the richest person in the world, the proprietor of one of the largest, most influential global companies, and a person of global political clout—are suspended; in the lingo of the digital discourse, Bill Gates is simply a nerd. And in the Microsoft universe he is certainly the nerd of all nerds.

In a world built on rational, ordered, logical lines of code, the nerd is (seemingly the antihero, but in reality) *the* cultural hero, and being labeled as one is invariably a superlative (recall "nerd wealth"). To be a nerd is first and foremost a pronouncement of computer skills. Fondly referring to his brother, the narrator of "Microserfs" acknowledges, "He was way better with computers than I was. He was way nerdier than me" (Coupland 1994). He goes on to theorize the connection between a personality type and work; here the identity of the nerd is presented as intertwined with information work and technology: "Nerds," he says, "over-focus . . . it's precisely this ability to narrow-focus that makes them so good at code writing: one line at a time, one line in a strand of millions" (Coupland 1994). The digerati are not simply a class; they are a caste, based on a race of technologically savvy individuals, of nerds. As another passage points out, what started out as a derogatory label in the 1970s for the socially unfit, the "underdog," came to be hip and prestigious in the 1990s, not least of all because of the fact that "now nerds run the world!" (Coupland 1994).

The meritocratic ethos of the digerati, based on expertise in the dominant force of production, is adverse to—and even annuls—other forms of social classification, specifically those based on the relational position in the mode of production (such as *class*) or those based on ascription (such as racial, gender, or national *identity*). Tom Campbell sums up this ethos in reference to the culture of Silicon Valley, saying, "We believe in liberty. We're comparatively non-prejudicial—in the Valley, it's what you've done, not who you are. It's what you believe, not who your family is. We believe in the ability of the individual in a country that allows us to be free" (Bronson 2003).

CONCLUSION: FROM THE BUREAUCRAT TO THE DIGERATI

According to the digital discourse, with the integration of the traditional world of work into network technology, the boundaries between work life and personal life become blurred, and workspace and work time are intermingled with their private, personal counterparts. These novelties allow workers to bring their personal qualities of creativity and deep personal engagement to bear on their work activities and reeroticize the disenchanted world of (industrial) work. They also render the workplace flat, that is, decentralized and dehierarchized, eradicating hierarchies between workers on the one hand and managers and owners on the other. What comes in their place is a regime of meritocracy and professionalism based on one's passion for, and proficiency in, network technology. These narratives of network work signal a radical break from the narratives of work alienation in the bureaucratized organization. Network technology is situated at the pivotal axis of rendering work—the work process, work relations, and the workplace—more humane and more liberating for the worker.

At a more macro level, the narratives concerning network work also signal a shift from a Fordist discourse of *class* to a post-Fordist discourse of *networks*. In the discourse of networks, individuals are construed as autonomous nodes and defined by their connections to other nodes in the network; the social world is seen as a flat, decentralized sphere of ever-flowing, multiple, and ad hoc assemblages. This is a stark contrast to the Fordist conception of the social sphere as consisting of a hierarchized, stable, and category-defined arena. While the discourse of class stresses structural power relations, the discourse of networks is devoid of such a conception of power and instead associates power with the characteristics of autonomous nodes (i.e., power resulting from ingenuity, techiness, nerdiness, and entrepreneurship). The digerati, the new protagonists of network work, can no longer be defined by their position in relation to production (i.e., by their class) but are defined in purely meritocratic terms, or in terms of their relationships with the key means of production, network technology (i.e., their technical expertise). In the same vein, the central mode of social action in the network discourse is that of cooperation rather than struggle or competition, which characterizes the discourse on class. The notion of network work and its grounding in a technological reality allow for the substitution of a hierarchical, competitive, and antagonistic model of class by a dehierarchized, cooperative, agreeable, and inherently inclusive model of networks.

The digital discourse on network work is understood in this book as an ideational structure that supports the transformations of work under post-Fordism. Specifically, the narratives of network work should be seen as an articulation—through mediation and naturalization of a technologistic language—of a response to the critique of Fordist capitalism centered on industrial technology and built on authority, hierarchy and centralization. The digital discourse tells of a radical break from industrial-age order, and network work and the digerati are seen as the complete opposite (and in fact rejection) of their industrial counterparts. The work process is seen as a site of emancipation and personal freedom rather than of alienation, and network technology is rendered an axis through which social tensions are alleviated.

DEHIERARCHIZATION AS CONTROL

While capitalism, and the market in particular, is said to thrive on spontaneous order—the decentralized mechanism that self-regulates the actions of individuals—corporations, the staple of capitalist markets, have

grown throughout most of the history of capitalism to rely on centralized, planned order. The technology of the bureaucratic organization was
imported by capitalist corporations and the modern state from the military. It is there, specifically in Bismarck's Prussian military, that an organization based on impersonal ranks (or offices), hierarchy, clear orders, and
record keeping was implemented on a large scale (Sennet 2006, 18–37),
bringing about the "militarization of civil society" (Sennet 2006, 20).
Planning, centralization, and control were seen as indispensable tools for
capitalist organizations to achieve economic rationality. The bureaucratic,
economic organization achieved better efficiency, control, and predictability through the construction of a hierarchical architecture and centralized control mechanisms.

In the last thirty years, however, corporations have gone through a
dramatic transformation in their mode of operation, from a model of
"vertical integration of departments within the same corporate structure"
to one of "vertical disintegration of production along a network of firms"
(Castells 1996, 158). Within the corporation, this has led to a paradigm
shift from vertical bureaucracies to the horizontal corporation (Castells
1996, 164), which Castells terms the "network enterprise" (see Castells
1996, chap. 3) and which is characterized by decentralization of production processes and dehierarchization of management practices.

However, as hierarchy went out the window, control was decentralized, and work processes flowed horizontally rather than vertically, the
underlying rationale remained very much the same. According to critical
perspectives on work and technology, network work was not simply an
effect of network technology but entailed a violent practice of reengi-
neering, "which was used to flatten organizational hierarchies, redefine
required skills, and introduce new technology" (Greenbaum 1995, 92).
Joan Greenbaum argues that the incentive for collapsing hierarchies in
the workplace was to increase the flexibility of organizations. Measures,
such as the elimination of titles, were meant to have "workers . . . assume
more functions and be more 'flexible'" (Greenbaum 1995, 100). Like
Sennet (2000), Greenbaum, too, argues that this demand to shift between
tasks and positions within organizations resulted in undermining workers' ability to construct long-term, stable, and meaningful careers (Greenbaum 1995, 101).

Another measure in the construction of the new network workplace
was the adoption of "the 'entrepreneurial' ideal—where workers would
fend for themselves" (Greenbaum 1995, 92). According to Greenbaum,
this meant that workers had to absorb the "anxieties" of a freer, more volatile market (Greenbaum 1995, 92). Even within the workplace, work was
privatized so that workers and teams often compete among themselves in

an "internal market" (Sennet 2006, 52). This new enterprise workplace (Greenbaum 1995), where workers were encouraged to think of themselves as entrepreneurs and act accordingly, also impacted the position of workers as a social class. Robert Reich, secretary of labor under President Clinton, coined the term "the anxious class" to account for the "millions of Americans who no longer count on having their jobs next year, or next month" (Uchitelle 1994, cited in Greenbaum 1995, 82). Simon Head adds that "the harsh and unstable work regime of the New Economy undermine[s] the security of employees and weaken[s] their bargaining power in the workplace" (Head 2003, xiii). A more decentralized and dehierarchized workplace, one that increases personal freedom and decreases alienation, is also a workplace that transfers risks from the company to individual and where working lives are more privatized and precarious.

Hence, behind the flattened and more humane workplace still lurks the dialectics of forces of production and relations of production presented at the beginning of this chapter. In a study on corporate culture in high-tech industries, Gideon Kunda (1992) concludes that the dehierarchized management style prevalent in this setting elicits more, not less, control over workers. These corporations, argues Kunda, employ a set of managerial policies "designed to minimize the use and deemphasize the significance of traditional bureaucratic control structures . . . and to elicit instead behavior consistent with cultural prescriptions. Thus, Tech's job-category and compensation systems . . . are grounded in the unchanging bedrock of traditional employment practices. However, they are complemented by additional policies specifically designed . . . [among other things] to remove some of the symbols, and realities, of status, supervision, and formal control" (Kunda 1992, 218–19). Formal control in the high-tech corporate environment, Kunda maintains, is substituted by a "softer" normative control (Kunda 1992).

TECHNOLOGY AND CLASS CONSCIOUSNESS

The digital discourse juxtaposes two sets of narratives: On the one hand, there are the emancipatory narratives that highlight the individual empowerment of the digerati resulting from the decentralization, dehierarchization, and the flattening out of relations within the workplace. On the other hand, there are the narratives that stress the extent to which the integration of these new characteristics into the capitalist corporation makes its workings more beneficial. For example, mobility increases the freedom of both workers and organizations by facilitating flexibility. The Archimedean point for both of these processes is network technology, which makes work not only more humane (from the point of view of the process) but also more rational (from the point of view of the end result of accumulation). This argument has been most eloquently articulated in the work of Daniel Bell.

Bell's theory regarding *The Coming of Post-Industrial Society* makes similar assertions regarding the depoliticized nature of technology and the corollary emergence of a meritocratic, classless society, one that is flat and networked. Bell's analysis essentially takes out the element of power from the conflictual framework traditionally applied to the analysis of the labor process by arguing that the emergence of information and technology as the prominent forces of production brings about a new social order that invalidates the classical Marxist analysis of society. According to Bell, Marx's concept of the "mode of production" is problematic and, in fact, conceals two *independent* variables: social relations and technological development (Marx's relations of production and forces of production, respectively). Contrary to Marx's fundamental assumption regarding dialectical and conflicting relationship between relations of production and forces of production, Bell argues that they do not necessarily correlate. The shift from slavery to feudalism to capitalism has a history independent of the shift from preindustrial to industrial and then to postindustrial society; after all, communist states were no less industrial than capitalist states (Bell 1999, xxix–xxx; see Figure 4.2. Also compare with Figure 4.1).

Bell, therefore, is able to depict a society where the most acute social tensions—between capital and labor—are alleviated in a way that is not only nonrevolutionary but where the triumph of capitalism does not drift toward a ruthless market society and the dictatorship of the capitalist class.

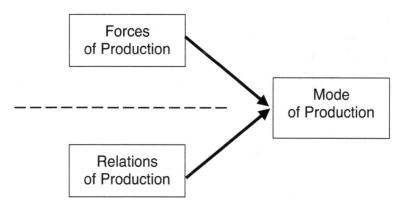

Figure 4.2. Bell's formulation of the mode of production as determined by two independent factors: productive forces and productive relations

Bell substitutes the rise to eminence of a new class of professionals and technocrats for the dichotomy of capitalists and proletariats; instead of the dichotomous alternatives of either a market society or a proletariat dictatorship Bell identifies a society, the axis of which is based on the new force of production, knowledge, information, and information technology. Hence, Bell is able to envision the coming of a social order in which technology replaces class struggle. Andrew Ross summarizes the thrust of Bell's argument as "a dream in which technology, rather than class conflict, would usher in a new realm of freedom for workers" (Ross 2003, 11).

What Bell and the digital discourse suggest is a technological overcoming of what Marx saw as the source of conflict in society: the dialectics of productive forces and productive relations. According to the digital discourse, these dialectic relations—and the type of economy and society they produce—are now being transformed by network technology. The social, this argument goes, is now predicated not on the tension between the two but on the triumphant powers of the forces of production in the form of information technology.

The most promising of these transformations, according to Bell, is the emergence of a technocratic elite (comprised of professionals at the forefront of the information and knowledge sectors) and the technical administration of the capitalist economy. The postindustrial society, he therefore predicts, is a society characterized by the substitution of class struggle by rationalization brought about by technical and technological means and championed by the technocratic elite, which has an inherent tendency to act rationally. The most decisive social ramification of the coming of postindustrialism is the subordination of the market economy to political and technical rationale. Further bureaucratization and technocratization of postindustrial society means that the defining social division is no longer based on property relations between those who own the means of production and those who do not, but "the bureaucratic and authority relations between those who have powers of decision and those who have not, in all kinds of organizations, political, economic, and social" (Bell 1999, 119).

As illustrated in this chapter, the identity of the digerati is constructed around the axis of the forces of production, while production relations are mostly ignored, downplayed, or plastered over. This reformulation helps reconceptualize technology as an axis of change and transcendence of social relations and a tool of personal and social emancipation rather than the alternative, mostly Marxist formula, where technology is always embedded with capitalist social relations and is therefore more likely to reproduce rather than transform them.

The digerati are the carriers of a new spirit, a new subjectivity, which embodies the core characteristics of a classless society. The digerati construct an (occupational) identity based on the separation of technology from social relations. Network technology is substituted for social relations of power as an axis of class composition, class structure, and, ultimately, a sense of identity. The digital discourse features a split between technology and power or, more theoretically, between forces of production and relations of production, with the latter receding into the background and the former constituting the central axis of the social sphere. In the context of this chapter, work, class, and identity are rearticulated as a derivative of technology, and power is taken out of the equation.

CRITIQUE IN THE CULTURE OF THE NEW CAPITALISM

This formulation of a classless society, where social power is determined not by ownership over the means of production but by technological dexterity, can partially explain the oddity of referring to the digerati entrepreneur with such a plebeian demeanor as we have already seen, as was the case with Andy Grove. But there is another facet of the digerati entrepreneur that must be elucidated: his contrarian, antisystemic, critical posture *vis-à-vis* the very social order of which he is integral and most significant part. To understand this oddity, it is important to point out here the historical sources of what I am calling in the book the "humanist critique." In the context of its historical sources the name that Boltanski and Chiapello (2005) are using—the "artistic critique"—is more telling. This type of critique of capitalism emerges from artistic and bohemian circles of Europe's nineteenth century and is closely linked with the values and aspirations of artists in that era: authenticity, self-expression, noninstrumentality, and critical distance from the prevailing social order, particularly from its most systemic structures: the market and the state.

This culture of critique, the negation of and resistance to prevailing bureaucratic and capitalist norms and to the "system" at large, has found its way into the spirit of networks, most noticeably in the presentation of the digerati, particularly the digerati entrepreneur. An antihero hero, a persona of resistance and negativity, a contrarian and revolutionary, this "great man" is at one and the same time a *pillar* of contemporary social structure and an emblem of its *critique*. Put differently, negativity and critique have become part of the culture of the new capitalism. More than that, they have become essential forces of production, that is, features of the counterculture that became part of the digital culture, a contrarian "cool" that is now part and parcel of work life for the informational worker. Writes Alan Liu in *The Laws of Cool* (2004), "Increasingly, knowledge work has no true recreational outside. Cool therefore arises *inside* the regime of knowledge work as what might be called an *intra*culture rather than a subculture or counterculture. Cool is an attitude or pose from within the belly of the beast, an effort to make one's very mode of inhabiting a cubicle express what in the 1960s would have been an 'alternative lifestyle' but now in the postindustrial 2000s is an alternative *workstyle*" (Liu 2004, 77–78). In the same vein, Richard Lloyd speaks of "the bohemian ethics and the spirit of flexibility" (Lloyd 2005, 235) as characterizing the new postindustrial social category of informational workers, Thomas Frank and Matt Weiland (1997) lament the commodification of dissent, and Naomi Klein (1999) has persuasively shown the co-optation of countercultural symbols by corporations in contemporary, informational capitalism.

The "artistic" undertones of critique and negation in the spirit of networks—the backbone of *The New Culture of Capitalism* (Sennett 2006)—renders the humanist critique also the "critical critique," at the heart of which is a requirement for critical distance from the system and from its effects (power, comfort, and privilege). It is this type of critical distance that critical theory attempted to advance in theory and in practice. And it is precisely this type of critique that we now see at the heart of the spirit of the new capitalism, "from within the belly of the beast" in Liu's words (Liu 2004, 77). Chapter 8 will ponder the significance of the incorporation of negation and critique into the spirit of network capitalism and will assess its viability as a substantive critical stance.

CHAPTER 5

NETWORK PRODUCTION

THE PREVIOUS CHAPTER LOOKED AT THE TRANSFORMATIONS IN THE world of work in the confines of the workplace. It asked how, according to the digital discourse, the old dynamics of work and structures of power in the industrial workplace have been transformed by their integration into network technology and how this has transformed the social structure in general. The current chapter ventures out of these boundaries and examines the digital discourse at the intersection of network technology and the process of production in general and the completely new dynamics that it brings about. According to the digital discourse, network technology does not simply change how companies do business and how production is carried out within the traditional workplace; it also makes a complete transformation of the very process of production. The chapter asks what, according to the digital discourse, happens to the process of production when it is integrated with network technology. Specifically, how are labor and production reshaped and redefined, and what are the broader social ramifications of these transformations? In short, what is network production?

According to the digital discourse, network technology allows for a more democratic, participatory, and inclusive mode of production. It reinvigorates the process of production, making it more humane and engaging for individuals by welcoming and harnessing human facets, such as authenticity, personal expression, and creativity, that have been excluded from production up to this time. In that context emerges the process of *prosumption*, a hybrid of production and consumption that renders both practices more engaging, participatory, and fulfilling; here also emerges the "prosumer" as a key social actor. Network production also facilitates the crystallization of emergent, self-governing, and self-regulating collaborative projects, such as open source. The intrinsic flexibility of networks allows for a more sophisticated utilization of resources such as play, joy, and free time that can be harnessed for wealth creation. In that respect, network production allows for a more rational process from a systemic point of view as well. The chapter presents key terms in the discourse on network production (e.g., prosumption, open source, flexibility, democracy, participation, interactivity, and so forth) and uncovers their hegemonic dimensions—that is, their fit into the operation of the new capitalism.

POWER TO THE PEOPLE:
DEMOCRATIZATION AND INDIVIDUAL EMPOWERMENT

According to the digital discourse, network production entails more than what managerial discourse calls "streamlining" (i.e., the rendering of preexisting production processes more smooth and efficient). While this quantitative progress no doubt takes place, it is the qualitative and revolutionary leap brought about by network technology that is of greater consequence. Network production is production made not simply better, but different. It entails a new *gestalt*, the full scope of which could not have been fully materialized and grasped only a few years back, not even through much of the history of digital technology. While the history of the digital discourse, and *Wired* as its epitome, is relatively short, one can nevertheless delineate a trend whereby *networks* are increasingly seen as the centerpiece of the digital revolution, displacing both *information* and *communication*.

At first, the information revolution was seen in the digital discourse as being anchored in the ability to translate the world into binary code—to informationalize it and hence subject it to computational manipulations. The technological centerpiece of that discourse was therefore the stand-alone computer—a machine of great autonomous computational ability (exemplified in mainframes, artificial intelligence, robots, etc.)—and applications such as Word and Excel. Then the interconnection of individual computers over communication lines and the ability to transfer information and eradicate time and space was seen as the hallmark of the information revolution. The technological emblem of communication was the infrastructure of the Internet and applications such as e-mail clients and Web browsers. Most recently, the network—not simply as a channel for the transfer of information between nodes, but as a site unto itself—has displaced both information and communication as the central locus of the information revolution. While here, too, the Internet is the centerpiece technology, this latest phase is distinguished in the digital discourse from the previous one with such terms as "second-generation Web" (Anderson 2006), or "Web 2.0" (Levy 2005), referring to network-based applications such as Wiki (which enables Wikipedia), Google, blogs, and MySpace, which are based on activation and production of content by users. Google, for example, can be seen as an application for the accumulation and organization of information regarding users' choices; MySpace and YouTube are only as vibrant and dynamic as their users. Hence, a recent article in *Wired* identifies the "death" of the PC as one of the leading trends in digital culture and the digital economy in particular and declares "Desktop R.I.P." (Tanz 2007). The PC, the article announces, is being replaced by "the cloud" (Tanz 2007)—that is, the network.

The potential of the network to bring about a new and revolutionary mode of production is explored in an opening article of a special issue of *Wired* for the tenth anniversary of the Web, a date marked by the initial public offering (IPO) of Netscape (Kelly 2005).[1] For Kevin Kelly, the author of the article, the abstract, ideal-typical "network" receives a concrete technological articulation in the form of the Web. Kelly asks two questions: What is the new mode of production brought about by the Web? And what is the major social meaning of that? The title and subtitle of his article make the following statement: "We are the Web. The Netscape IPO wasn't really about dot-commerce. At its heart was a new cultural force based on mass collaboration. Blogs, Wikipedia, open source, peer-to-peer—behold the power of the people" (Kelly 2005).

Contrary to the commercial-oriented framework by which the Web was understood in the 1990s (testified by such terms as the "dot-com boom," or the "new economy"), Kelly offers a completely different narrative, one that takes production out of its strictly economic context. Network production, he suggests, is about the empowerment of "the people." This proposition brings two questions to mind: Who are the people? And what is the new power they are endowed with thanks to network technology? According to Kelly, the most exciting potential of the network is in allowing individuals to come together and unleash a powerful, creative, and democratic force of production; the new mode of production is based on the collaborative work of individuals. The Web allows for "a new kind of participation that has since developed into an emerging culture based on sharing" (Kelly 2005). That means that production—here Kelly focuses mostly on cultural production—has been democratized: "Suddenly it became clear that ordinary people could create material anyone with a connection could view. The burgeoning online audience no longer needed ABC for content."

This democratization further means that professionals and experts no longer dominate the processes of production; now Web "users" (in Kelly's terms) do most of the work. In managing auctions, for example, "they photograph, catalog, post, and manage their own auctions." This collaborative mode empowers individuals and is democratizing in another important way: not only does it allow lay individuals to engage in production;

such collaborative organization requires nobody to be in charge, thereby flattening out hierarchies and eliminating filters set by power structures. Again, in the case of auction sites, individual nodes in the network mode of production (i.e., individual users) "police themselves . . . The chief method of ensuring fairness is a system of user-generated ratings. These billion feedback comments can work wonders" (Kelly 2005).

The grassroots character afforded by the Web empowers "the people"—the little men, the individuals—rather than (or better yet *vis-à-vis*) large institutions such as states and corporations. Therefore, this new network production entails, according to Kelly, a radical break from the old economy and even from some of the fundamental tenets of capitalism. First and foremost, the new world of production, brought about by the Web, is "manufactured by users, not corporate interests." What's more, users' behavior does not follow the capitalist rationality of profit maximization. Speaking of blogs, Kelly points out that "these user-created channels make no sense economically." The "time, energy, and resources" involved in bringing about such products are "coming from . . . the audience" rather than from the traditional wage laborer. In addition, the fruits of such labor are also increasingly noncommercial. All of these examples, says Kelly, crystallize into a "gift economy," where "all [is] given away for free" (Kelly 2005).

The network mode of production undermines traditional conceptions of production and consumption. It "permits easy modification and reuse, and thus promotes consumers into producers," transforming many endeavors "from spectator art to participatory democracy." The Web allows the transformation of passive audiences into content providers and of consumers into producers; it "assumes participation, not mere consumption," and ushers in a "great shift from audience to participants." The Web allows the sort of engagement with the world that follows the logic of *interaction*. It involves "deep enthusiasm for making things, for interacting more deeply than just choosing options . . . This impulse for participation has upended the economy and is steadily turning the sphere of social networking—smart mobs, hive minds, and collaborative action—into the main event" (Kelly 2005).

Network production, according to this, is not only postcapitalist; by breaking the dichotomy of production/consumption and producer/consumer, it is postproletariat and postconsumerist as well. Thanks to the Web, creativity, enthusiasm, cooperation, and personal expression are reintroduced into the spheres of production and consumption from which they have been traditionally excluded and revitalize them. With network production, individuals are no longer simply peons of a consumerist mass society; they regain their agency and independence *through* production and consumption. Kelly predicts that the Web will bring further democratization by toppling a mass homogenized society and providing greater opportunities for individuals to be creative and self-expressive. "The Web," he says, "continues to evolve from a world ruled by mass media and mass audiences to one ruled by messy media and messy participation . . . In the near future, everyone alive will (on average) write a song, author a book, make a video, craft a weblog, and code a program" (Kelly 2005).

In such mass popularization and democratization of production on the Web, no one, says Kelly, would be a consumer. Such passive categories as consumers and audiences would simply vanish, because what matters on the Web is not audience; "what matters is the network of social creation, the community of collaborative interaction that futurist Alvin Toffler called prosumption . . . Prosumers produce and consume at once" (Kelly 2005). Power is delivered to "the people" by the democratization and decentralization afforded by network technology, which in turn allows participation and collaboration. As the means of production are democratized, so is the mode of production and social relations in general.[2]

But the collaborative nature of network production not only empowers individuals, enabling a more democratic participation and a more engaged and satisfying experience in production. Kelly's insistence that "the sphere of social networking . . . [is] the main event" of the network society is backed by another set of arguments. The end result of these collaborative efforts—the *what* that is being produced (i.e., the network as a *product*) also turns out to be superior to anything we have experienced in the past. The network as an end result functions as a depository of knowledge, furnishing humans with a newfound and a quasi-transcendental view of the world: "This view is spookily godlike. You can switch your gaze of a spot in the world from map to satellite to 3-D just by clicking. Recall the past? It's there. Or listen to the daily complaints and travails of almost anyone who blogs (and doesn't everyone?) I doubt angels have a better view of humanity" (Kelly 2005).

The Web with a "scope . . . [that] is hard to fathom" (Kelly 2005), expands our ability to know the world and ourselves like never before. But more than merely a passive archive of knowledge, the network becomes an active living brain, a "collective intelligence," in the terms of Pierre Levy (1997). According to Kelly, "The ways of participation unleashed by hyperlinks are creating a new type of thinking—part human and part machine—found nowhere else on the planet or in history" (2005).

Such intelligence can only come about as a result of the collaborative efforts of individuals, since it is based precisely on the democratic and decentralized process of network production that allows individual users to collaborate and work together as a hive. In network production, the seemingly consumerist action of each individual also alters the nature of the product, making it better. For example, Kelly explains that "Google turns traffic and link patterns generated by 2 billion searches a month into the organizing intelligence for a new economy" (Kelly 2005). Another example is Wikipedia, the online collaborative encyclopedia: "Each time we forge a link between words [by clicking a hyperlink] we teach it an idea . . . Wikipedia encourages its citizen authors to link each fact in an article to a reference citation. Over time, a Wikipedia article becomes totally underlined as ideas are cross-referenced" (Kelly 2005).

Kelly's enthusiasm is predicated on the network as *both a process of production and as its end result.* On both accounts, the network is superior to previous forms of the organization of production. As a process, it is more democratic, more engaging and satisfying to individuals, and more empowering to them. As an end result, or product—either as collective intelligence or simply "culture"—it is able to deliver a superior product compared with previous modes of production. In short, Kelly is able to connect an ethics of democracy, individual autonomy and empowerment, and ungovernability, together with the traditional ethics of capitalism, focused on productivity and the quality of the end product.

Let us now return to Kelly's proclamation at the top of his article—"behold the power of the people"—and to the questions I have posed about it: Who are "the people," and what is the kind of power that network technology gives them? And, what does it tell us about the spirit of networks? The double-spread image that precedes Kelly's article encapsulates some revealing answers.[3]

The image is comprised of sixty-three portraits of individuals. What can we tell about these people? They seem to differ in almost every respect: age, habitus, class, race, and lifestyle. In fact, their common denominator could only be inferred in the context of the article: they are all connected to the Web. They *are* the Web, as the article's title maintains—truly the "family of Web users." As the image reveals, "the people" of the digital discourse are not a monolithic group; the image is decidedly not a group photograph. "The people" of Kelly's Web are not copresent in the same place at the same time; they most likely do not know each other. They are facing us (or outward) rather than each other (or inward), and each is surrounded by a clear boundary from the others, leaving each in his or her own "bubble." "The people" are autonomous units, the interrelatedness of which is only possible through their interconnection by network technology, or the Web. "The people" are individuals who are able to come together while retaining their individuality, a loosely connected network of individuals who assemble ad hoc; they are minimally related, but their relation is dependent upon the glue of network technology. "The people" are the new army of network production.

OPEN SOURCE:
FUSING RATIONALIZATION AND EMANCIPATION

One of the technological emblems of network production as contributing to the power of "the people" is the open source architecture. The term "open source" referred originally to software products but came in the digital discourse to epitomize and indeed stand for the revolutionary potentialities entailed by the broader phenomenon of network production. Its broader political and social ramifications are frequently referred to in the context of the "open source movement," or "open source culture" (see, for example, the entry "open source" in Wikipedia). Let me start with a brief explanation of what open source is. Computer programs, such as Web browsers, word processors, or games, are written in a specialized programming language as lines of code. Once written, this collection of command lines, or source code, is habitually sealed (both technically and legally) so that users can run the program and use it as an application but have no access to its source code, its underlying programming text. From the point of view of software companies, this helps protect their proprietary rights, ensuring that their investment in writing the source code yields profit. From the point of view of users—specifically hackers and other tinkerers—the implication is that the program is a sealed product that cannot be altered and modified to their individual needs, nor can parts of it be copied and used as a basis for further development of software products. The open source movement reverses this situation by making the source code of the software technically and legally open; users have free access to it and can modify it to their needs or use parts of the code in order to build new software products. This, of course, has far-reaching implications for the process of production, as well as for the legal status of the end product.

One of the original open source projects, as well as arguably one of the most famous and successful ones, is Linux. Linux is an open source operating system inaugurated in 1994 by Linus Torvalds as an alternative to the existing proprietary operating systems of Microsoft and Apple. Torvalds wrote the kernel of the operating system, posted its source code on the Web, and invited other programmers to tinker with it and improve on it. It has since sprawled to become a viable (though still weak) competitor to the mainstream, proprietary operating systems and is now used by more than 18 million people around the world and in varied machines, from cell phones to servers (Rivlin 2003).

Wired's cover story of Linus Torvalds and Linux demarcates the two issues around which the discourse on open source revolves: collaboration (the subtitle of the article hails Linux as "the biggest collaborative project in history") and property rights (the title hails Torvalds as "leader of the free world"; Rivlin 2003). These two concerns correspond to the more general template of discussion I identified previously: networks as a production process and networks as a product. On both concerns, the digital discourse suggests open source resolves and transcends the problematics of the old capitalism. It offers a more democratized and decentralized production process and a product that empowers individuals and individualism by being nonproprietary. Linux, and open source in general, epitomizes the new mode of network production based on collaboration, a template for the transformation of the whole mode of production in contemporary society, and with it the transformation of society as a whole. The article on Linux in fact ends with a prediction that the open source model will be exported from software to other human activities and in turn "catalyze stagnant sectors of the economy—or, better yet, create new economic sectors" (Rivlin 2003).

This prediction is explored in another article, in which the author finds "open source everywhere" (Goetz 2003). The article discusses open source as a generalized template for social production. The revolutionary crux of open source as an epitome of network production is anchored, according to the digital discourse, in the new and improved mechanism for the division of labor that it facilitates: "Open source harnesses the distributive powers of the Internet, parcels the work out of thousands, and uses their piecework to build a better whole . . . It works like an ant colony, where the collective intelligence of the network supersedes any single contributor . . . In 2003, the method is proving to be as broadly effective—and, yes, as revolutionary—a means of production as the assembly line was a century ago" (Goetz 2003).

This framework locates network production on the continuum of the process of the rationalization of production. What Adam Smith's division of labor and Friedrich Taylor's scientific management are for the wake and the zenith of the industrial era, respectively, so is the digital discourse's open source for the emerging network revolution: all allow for the creation of a superior product, or "a better whole," by implementing ever more sophisticated means of coordination between fragmented and differentiated parts of the labor process.

However, as this framework suggests, open source is not simply a continuation of the rationalization of production. It also constitutes a radical break in that, for the first time since the industrial revolution, the rationalization of the work process through technology helps workers realize their freedom rather than contribute to their alienation. For the first time, this rationalization process is not only harmless to its practitioners, but, quite the contrary, it benefits them. Put differently, in contrast to previous advances in the mode of production, open source responds not only to the *rationalization* demands of production but also to the *emancipatory* demands of workers. Reiterating Kelly's discourse on the Web, the discourse on open source also fuses democracy and capitalism together, heralding the emergence of a more democratic and emancipatory form of rationality:

> Open source embodies an ethos as fruitful and resilient as the closed capitalism Bill Gates represents: the spirit of democratic solutions to daunting problems. It's the creed of Emerson, who preached independent initiative and advocated a "creative economy." It's the philosophy of William James, whose pragmatism dictated that "ideals ought to aim at the transformation of reality." It's the science of Frederick Taylor, who proved that distributing work could exponentially boost productivity and replace "suspicious watchfulness" with "mutual confidence." It's the logic of Adam Smith, whose notion of "enlightened self-interest" among workers neatly presages the primary motivation for many open source collaborators. (Goetz 2003)

Open source represents a new kind of capitalism, characterized as open (as opposed to "the closed capitalism Bill Gates represents"), democratic, conducive to individual independence and creativity, pragmatic, decentralized, collaborative, and distributive. The brilliance of open source, according to this narrative, lies *neither* in just being "for the people" *nor* in just being a new and powerful source of wealth creation for corporations; rather, its revolutionary potential lies precisely in being *both*—in bridging and even transcending contradictory interests. That is set in contrast to the old capitalism, which responded only to its systemic demands and neglected to respond to the lifeworld demands of freedom and creativity.

The digital discourse sees open source not as "anticommercial or anticorporate" (Goetz 2003), as some commentators suggest, but rather as a new and improved form of capitalism that is more democratic and emancipatory. As the next passage (which could easily pass as a quote from *The Communist Manifesto*) exclaims, open source offers a superior capitalist tool and *also* holds the promise of workers' emancipation, a truly universal instrument benefiting both capitalists and workers: "While the assembly line accelerated the pace of production, it also embedded workers more deeply into the corporate manufacturing machine. Indeed, that was the big innovation of the 20th-century factory: The machines, rather than the workers, drove production. With open source, the people are back in charge. Through distributed collaboration, a multitude of workers can tackle a problem, all at once. The speed is even greater—but so is the freedom. It's a cottage industry on Internet time" (Goetz 2003).

The network mode of production offers a way of responding to two demands that are at the heart of modernity: economic rationality and human emancipation. Not only were the two demands historically thought to be at odds with each other; moreover, the zenith of industrial society, with its centralized, mechanical technology and its organizational corollaries, came to epitomize the suppression of the demands for emancipation through work and technology. But now network technology is construed as the means by which the two demands can be met. Moreover, it is met not by a political, externally imposed compromise between contrasting interests (which was ultimately what the welfare state and Fordism was all about) but rather by their complete fusion or enmeshment through the internal rationale of network technology and its corollary, network production: "With open source, you've got the first real industrial model that stems from the technology itself, rather than simply incorporating it" (Goetz 2003). Network production absorbs both rationalization and emancipation and resolves their hitherto contradictory (and crisis-laden) tendencies.

According to the digital discourse, network technology changes the balance of power between managers and capitalists on the one hand and worker and innovators on the other. In contrast with the assembly line, network technology puts the premium on workers—their skills, creativity, satisfaction, and their ability to control the process of production. In sociological terms, it allows for a process of reskilling after two-and-a-half centuries of deskilling (which many of the critics of technology—from the Luddites to Braverman, Noble, and Aronowitz—have been lamenting). And—echoing the language of Kelly—it puts "the people" back in charge (Goetz 2003).

CROWDSOURCING: REDRAWING THE LABOR LINE

The narratives of the digital discourse presented thus far assert the emancipatory effect that network production has on individuals, how it improves the rationalization of the process of production, and perhaps most important, how both emancipation and rationalization come to complement each other through network technology. The historically contradictory demands of capital, industry, efficiency, competition, and productivity on the one hand and the demands of labor, satisfaction, creativity, cooperation, and personal fulfillment and authenticity on the other hand come into alignment with the introduction of network production and, as a result, persistent tensions that characterized industrial production are resolved. But what are the terms of this resolution? If capital and labor align along a shared line, what is the shape of this line, and what is the topography of the terrain it traverses? To uncover some of those terms of truce between the demands for accumulation on the one hand and emancipation on the other hand, I will examine a key innovative technique of network production: crowdsourcing. The practice is relatively new and marginal in quantitative terms (i.e., it is not a dominant mode of production in contemporary capitalism). But because it is such an exemplary case of network production, the discourse on crowdsourcing helps uncover the hegemonic dimensions of the digital discourse in general. I will focus on two articles in *Wired* that report on crowdsourcing.

The term "crowdsourcing" is a play on the more familiar term "outsourcing." The first article heralds "The Rise of Crowdsourcing" (Howe 2006), and the subtitle reads, "Remember outsourcing? Sending jobs to India and China is so 2003. The new pool of cheap labor: everyday people using their spare cycles to create content, solve problems, even do corporate R&D" (Howe 2006).

Thus, the article anchors the rationale of network production in the ability to tap into a "pool of cheap labor" and mobilize it for one's production needs. This statement makes a distinction between outsourcing and crowdsourcing but at the same time uncovers how similar they are: in outsourcing, cheap labor is based on differentials in space; with crowdsourcing, cheap labor is based on differentials in time. The new source of cheap labor, this statement suggests, is predicated and conditioned on time that is deemed *unproductive*—"spare cycles" in the language of the article—that is, time that is not easily translatable to money and capital.

But what is the mechanism, or "trick," by which productive forces, harnessed for the project of capital accumulation, can nevertheless be conceived as unproductive (as "spare cycles") and hence be partially or wholly unpaid? Presumably the trick is technological—this is made possible by network technology. But the trick is also discursive: the trick is that this labor is construed in the digital discourse to be done by "everyday people," rather than by professional, tenured, "in-house" workers. In other words, it is labor done by amateurs during their free time as an expression of personal interest and passion, not labor done by professionals as an expression of the more mundane attempt to make a living. No wonder that the teaser for the article on the front cover of *Wired* proclaims, "Crowdsourcing: A billion *amateurs* want your job" (Cover 2006, emphasis mine). The threat is clear: one's position in the work force is threatened not so much by other workers but by a new class of workers: a new reserve army of (this time around) amateurs.

Network technology is not only at the heart of a revolution in the mode of production but also at the heart of remapping the field of "socially necessary labor" in Marx's terms (1990, 129), more specifially redrawing the lines between paid and unpaid labor. And to a large extent crowdsourcing, as a new discursive category of production, entails the transference of parts of the productive process from *paid, organized, professional* labor force to *unpaid, atomized, amateurish* enthusiasts. Underlying the digital discourse on crowdsourcing is the assumption that, because of network technology, more socially necessary labor can now be done without or with very little compensation: "The open source software movement proved that a network of passionate, geeky volunteers could write code just as well as the highly paid developers at Microsoft or Sun Microsystems. Wikipedia showed that the model could be used to create a sprawling and surprisingly comprehensive online encyclopedia. And companies like eBay and MySpace have built profitable businesses that couldn't exist without the contributions of users" (Howe 2006).

An increasingly large chunk of the new economy, according to this source, is built—indeed conditioned—on labor that is not compensated or, as we will see below, involves new, more precarious, and partial modes of compensation. In Marxist terms, this is a case of superexploitation, where the *whole* value produced is in fact surplus value. And these new relations between capital and labor, according to the digital discourse, have been a trend ever since the introduction of network technology, with companies getting ever more sophisticated at extracting profit with minimal or no monetary return to workers: "All these companies grew up in the Internet age and were designed to take advantage of the networked world. But now the productive potential of millions of plugged-in enthusiasts is attracting the attention of old-line businesses, too. For the last decade or so, companies have been looking overseas, to India and China, for cheap labor. But now it doesn't matter where the laborers are—they might be down the block, they might be in Indonesia—as long as they are connected to the network" (Howe 2006).

The network is seen in the digital discourse as a new repository for cheap labor, a mechanism for extracting hitherto unexploited resources. But looked at from the point of view of individuals, it is also a boon for small-scale producers: "Hobbyists, part-timers, and dabblers suddenly have a market for their efforts, as smart companies in industries as disparate as pharmaceuticals and television discover ways to tap the latent talent of the crowd. The labor isn't always free, but it costs a lot less than paying traditional employees. It's not outsourcing; it's crowdsourcing" (Howe 2006). The final sentence highlights the fact that crowdsourcing is ultimately interpreted in the digital discourse as a new source of profitability. In that sense, crowdsourcing offers not a break but a continuation of the long history of capitalist exploitation, based on increasing levels of surplus value (rather than simply increased levels of productivity). Outsourcing and crowdsourcing are equated, since they are both manifestations of what network technology can offer in ever more sophisticated manners: cheap labor and increased exploitation. In the uncritical, even triumphant language of the digital discourse, this is not seen as exploitation at all because labor is done by "amateurs," it is "latent," and it is done outside of productive time.

Network production entails a discursive redefinition of workers, with two key analytical innovations. The first is the dissociation of work (i.e., the investment of labor power in production) from employment (i.e., the institutional arrangement of livelihood, or the embeddedness of individuals within the social arrangement of production). The notion of "workers," in fact, becomes an unstable signifier standing for hobbyists and dabblers. By extension, this redefinition also legitimates new modes of employment. This is of course not only an innovation but also a continuation of a long process of the rationalization and automation of production where labor and skills are increasingly objectified and are hence dissociated from the laborer (Aronowitz and DiFazio 1994; Braverman 1974; Robins and Webster 1999).

The second analytical innovation is the construction of workers as "prosumers," an amalgam of producers and consumers. This renders the distinction between labor and capital obsolete and allows equating big corporations with small-time hobbyists; analytically, both are seen as homologous units of prosumption that can now meet in the digital marketplace and trade directly as equals. Moreover, with the redefinition of workers as prosumers, the category of "worker" as a distinct category is eliminated and instead the worker emerges as entrepreneur, an attribute I discussed in the previous chapter (Greenbaum 1995, 92; Huws 2003; Sennet 2006, 52).

It is through these analytical lenses that we can comprehend the astounding reference to the talent and work power of the crowd as "latent" (Howe 2006). According to this framework, only network technology is able to uncover and mobilize this talent and work power into productive results and, of course, with the condition that this talent is not paid or poorly paid. We must ask, then, why is it latent, and why can it only be uncovered and mobilized with network technology? Some answers can be glimpsed from some of the case studies of crowdsourcing explored in the article.

TOPPLING THE BASTION OF PROFESSIONALISM

The first story in Howe's article is told from the point of view of a consumer, Claudia Menashe, who was looking to buy photographs for an exhibition she organized. She contacted Mark Harmel, a professional photographer who offered what is known in the industry as "stock photos" (i.e., existing images) for about $150 each. Just before the deal was struck, Menashe stumbled upon iStockphoto.com, a stock photo Web site where she could find images for about $1 each. How, the reporter asks, could this site "undercut Harmel by more than 99 percent? . . . By creating a marketplace for the work of amateur photographers, homemakers, students, engineers, [and] dancers" (Howe 2006). Notwithstanding the harsh consequences for Harmel, the professional photographer, this is an upbeat story about the democratizing nature of network production, where everyone can "make it," since thanks to network technology, the bar for entering into the marketplace has been lowered like never before, and now "hobbyists, part-timers, and dabblers suddenly have a market for their efforts" (Howe 2006). This has completely transformed the business of photography: "Professional-grade cameras now cost less than $1,000. With a computer and a copy of Photoshop, even entry-level enthusiasts can create photographs rivaling those of professionals like Harmel. Add the Internet and powerful search technology, and sharing these images with the world becomes simple" (Howe 2006).

And so, "the product Harmel offers is no longer scarce." Network technology, according to the digital discourse, topples the bastion of what is by definition an exclusionary category: professionalism. It undercuts its guildlike privileges and instead empowers "the people" in two ways: hardware and software become cheap enough for nonprofessionals to acquire and a connection to the Internet renders virtually anyone a viable participant in the marketplace. "Technological advances in everything from product design software to digital video cameras are breaking down the cost barriers that once separated amateurs from professional" (Howe 2006); technology democratizes and popularizes economic opportunities. The boundaries between professionals and "enthusiasts" are blurred, since both have equal access to both the equipment and the social network needed to make and sell their products. Professionals and hobbyists are, of course, different in one crucial way: since the latter do not consider this activity their primary source of livelihood, they are ready to give away their products for free or very cheaply. These new circumstances create a market heaven: consumers are able to buy more cheaply, and amateurs are able to enter into the loop of production and creativity and offer the fruits of their passion. The only "loser" in this particular story about crowdsourcing seems to be Harmel, for whom photography is a profession and a career.

If network production undercuts professionals in favor of the masses of producers and consumers, one expects not only individual photographers would be harmed economically from a crowdsourcing initiative like iStockphoto but other actors would be harmed as well, such as big commercial agencies of stock photography. The reality, however, proves otherwise. These companies have been adapting to the new realities of network production by various measures. For example, Getty Images, the largest stock photography agency in the world, overcame the threat of economic damage by purchasing iStockphoto for $50 million in February 2006. Getty's CEO explains the move: "If someone's going to cannibalize your business, better it be one of your businesses" (Howe 2006).

Indeed, large companies, staples of the old capitalism, have been adapting to the new realities of network production by incorporating the new spirit of networks into their operating scheme. This, in turn, radically transforms the nature of production and indeed social relations as a whole. Another case study in the article focuses on the VH1 television channel hit show *Web Junk 20*, a compilation of videos shot by amateurs and posted on the Web. According to the article, this is not a solitary example; both network and cable television channels increasingly rely on crowdsourcing as a new mode of producing content on television. These channels include Bravo, USA Network, E!, and NBC. For example, more than 30 percent of Current TV's programming consists of materials submitted by viewers (Howe 2006).

Here, a somewhat different facet of prosumption is revealed, as the practice is presented from the point of view of large media corporations, who are able to take advantage of the crowd. As the article points out, as prices for the professional video clips that VH1 had been using rose, Michael Hirschorn, an executive with the channel, realized he could use "the growth of video on the Internet . . . and build a show around it." In a similar move to that made by Getty Images, Viacom—VH1's parent company—bought iFilm.com, a Web archive of video clips and VH1's main source for videos, for $49 million.

The new productive model offers obvious financial gains: "The model's most winning quality, as Hirschorn readily admits, is that it's 'incredibly cheap'—cheaper than almost anything else on television . . . Hirschorn thinks the crowd will be crucial component of TV 2.0. 'I can imagine a time when all of our shows will have a user-generated component'" (Howe 2006).

Notwithstanding the obvious economic sense of this scheme, the reporter nevertheless warily raises the question of whether "the crowd [can] produce enough content to support an array of shows over many years." Hirschorn readily puts his trust in the network and the quality of the pool of free labor it musters. He predicts that "as user-generated TV matures, the users will become more proficient and the networks better at ferreting out the best of the best." The article provides support for this optimism: a Pew survey found that 57 percent of twelve- to seventeen-year-olds online (twelve million individuals) are creating content and posting it on the Web. Moreover, in the case of iFilm.com, the crowd plays another important role in the process of television production: "Because iFilm already ranks videos by popularity, the service came with an infrastructure for separating the gold from the god-awful." The crowd in this case serves also as an editorial board. In this model, "viewers serve as both talent pool and jury." Hirschorn, therefore, predicts that "the model will succeed" (Howe 2006). In an ironic and historical turn of events, corporations such as VH1 are now vowing for a process of *reskilling*. They hope for "workers" to be more talented and skillful because crowdsourcing allows the company to meet them as producers in the marketplace, rather than as workers, and purchase strictly their products—that is, the end result of their labor—rather than their labor power and labor time. I will get back to that crucial point below.

The prime significance of crowdsourcing, according to the article, is not only economic. It is also not necessarily significant in terms of the cultural quality of the end products; most of these videos, the article admits, are vulgar and "stupid." What's important is the moral significance of crowdsourcing. The article suggests that crowdsourcing brings about a "new, democratic age of entertainment by the masses, for the masses" (Howe 2006). This populist ethos is also embedded in the names of many of the Web sites that thrive on network production. The formula "i + [content type]" (e.g., iStockphoto, iFilm) speaks to the empowerment of the "I"—the individual, amateur, unprofessional, "the people," the masses—who can nevertheless shoot his or her own film and distribute it over the Internet. The same goes for the name of the video-sharing site YouTube, where the second person "You" refers to each individual user, or the social networking site MySpace, which again speaks of a personal space. The irony in this kind of allusion to the individual and "the people" is that many of these Web sites—which originally emerged as grassroots operations—are now owned by huge media corporations: Getty, Viacom, Google, and NewsCorp.

CROWDS AND ENTHUSIASTS

Network production allows for a significant reduction in production costs by creating a business model based on the labor power of uncompensated and undercompensated laborers. But more than that, network production seems to be dependent on a *new social compact*, where "the people" receive more opportunities to engage their skills, creativity, hobbies, and passions and corporations receive hitherto unexploited sources for capital accumulation. On the other side of the coin, "the people" give up stable compensation schemes embedded in institutionalized employment and corporations surrender their claims for authority, professionalism, and control over content. The full scope and significance of this new social compact is unraveled in another case study of crowdsourcing, which goes to the heart of the new capitalism: the use of network production for research and development (R&D).

Network production can not only help mobilize the crowd in general but also separate the wheat from the chaff and get the best out of the crowd of highly skilled and professional individuals. "Not everyone in the crowd wants to make silly videos," posits the article. "Some have the kind of scientific talent and expertise that corporate America is now finding a way to tap" (Howe 2006). The underlying assumption in this statement, and indeed throughout the story, is that the old way of tapping into people's talent and labor power—by employing them—fell out of favor and that network production offers new ways to do it; indeed, network technology facilitates new *relations* of production.

The story focuses on one such professional—a retired physicist named Ed Melcarek—and one such mechanism for tapping into talent—the Inno-Centive Web site. Companies, the article says, are restructuring their R&D departments in accordance with crowdsourcing: "Forward-thinking companies are changing the face of R&D. Exit the white lab coats; enter Melcarek—one of over 90,000 'solvers' who make up the network of scientists on InnoCentive, the research world's version of iStockphoto" (Howe 2006).

The shift is from a standing army of scientists—tenured, employed full-time, and receiving benefits—to a reserve army in the form of a net-worked crowd. The reference to the "white lab coats" is telling: a staple, perhaps, of scientists' habitus (like a physician's stethoscope or a profes-sor's book under the armpit—a tool that is at the same time a profes-sional signifier) but also an allusion (or a connotation, in Barthes's terms) to a much deeper set of meanings. A white lab coat is the "uniform" of the worker; it stands for tenure, for being an integral part of a greater system—a peon perhaps, but a *member* nonetheless, without which the structure will not survive. The white coat—like the white collar and blue collar—is the emblem of the bureaucratic organization of industrial society *par excellence*. The figurative "exit the white lab coats" is therefore literal as well: exit tenure, mass employment, working class, collective bargaining, and so forth.

In contrast to the previous examples of iStockphoto and iFilm, where big media corporations have colonized grassroots initiatives after the fact, in the case of InnoCentive the initiative originates at the center of power: "Pharmaceutical maker Eli Lilly funded InnoCentive's launch in 2001 as *a way to connect with brainpower outside the company*—people who could help develop drugs and speed them to market. From the outset, Inno-Centive threw open the doors to other firms eager to access the network's trove of *ad hoc experts*" (Howe 2006; emphasis mine).

These are not small-scale "start-ups" or not-for-profit initiatives, but well-established, rich, profitable companies. In addition to Eli Lilly, companies using InnoCentive include Boeing, DuPont, and Procter & Gamble.[4] As the article puts it, these ad hoc experts "attack problems that have stumped some of the best corporate scientists at Fortune 100 companies" (Howe 2006). These companies "post their most ornery scientific problems on InnoCentive's Web site; anyone on InnoCentive's network can take a shot at cracking them." The obvious reason why companies now "turn to the crowd [is] to help curb the rising cost of corporate research" (Howe 2006); companies offer merely $10,000 to $100,000 per solution.

But the economic benefit is only half the story, according to the article. The advantage that these companies see in crowdsourcing is the ability to mobilize the collective intelligence, or "smartness," of the Web—smartness that does not exist anywhere else—to solve complex R&D problems. In fact, even though not all individuals in the crowd of solvers are experts and professionals ("the solvers are not who you might expect. Many are hobbyists . . ."), collectively, the article argues, the intelligence of the network is superior: "The strength of a network like InnoCentive's is exactly the diversity of intellectual background" (Howe 2006).

Network production brings with it new modes of employment that are more flexible and precarious. And the digital discourse offers an explanation of them based on the inherent superiority of the network as a mechanism for the mobilization of dispersed skills. Hence, the legitimation for these new types of ad hoc employment schemes is based not only on *economic reasoning* (that this type of employment costs less and is more flexible and less burdening from the point of view of employers), and not only on *emancipatory reasoning* (that it allows more flexibility from the point of view of employees who would rather not be locked in commit-ted—but possibly unsatisfying and stultifying—working relations; more on that below), but also on a new *network reasoning*, according to which the network is a repository of superior intelligence. Through the network, companies are able to tap into the expertise and knowledge of many more individuals than they can ever employ directly, and in turn, they can achieve better, smarter results (materialized in the form of products). Recall Kelly's assertion that "no one is as smart as everyone" (Kelly 1998, 14). This is also very much in line with Friedrich Hayek's conception of market information as tacit and dispersed (discussed in Chapter 3), but in this case the formula is applied not simply to information about markets but to information in the sense of specialized, professional, academic, and practical knowledge and expertise.

Network production is presented as the condition for two trends, which are conceived as intertwined and codependent. On the one hand is the universal benevolence of a new superior smartness, which can lead—who knows?—to the development of a new life-saving drug by Eli Lilly. On the other hand is the new labor regime of loose, ad hoc, flexible employer-worker relations epitomized by such projects as InnoCentive. These two trends are constructed as intertwined; that is, in order to reap the benefits of this new network rationality we must accept the more flexible, precari-ous, and privatized regime of employment, since both are predicated on the characteristics of the network. Network technology has the ability to make production more rational, but to do that successfully, it is implied, workers must be redefined as autonomous nodes in the network that can be flexibly mobilized to a specific project and then let go (see Figure 5.1.).

To be more precise, according to the digital discourse, to achieve supe-rior smartness we must accept flexible employment, since both are predi-cated on the characteristics of the network, especially on "the strength

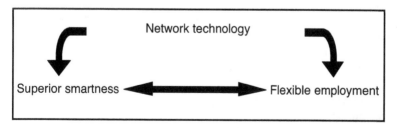

Figure 5.1.   The superior smartness of the network is conditioned, according to the digital discourse, by tracing the source of both the digital discourse and flexible employment schemes to network technology.

of weak ties," identified in the article as "a central tenet of network theory" (Howe 2006). Here, the notion of "weak ties" that the article borrows from sociologist Mark Granovetter is translated quite literally and unproblematically into the realm of production relations to mean weak ties between employers and employees, or between different nodes engaged in a collaborative project, as the digital discourse would have it.[5]

The linking of network technology to both a new labor regime and superior smartness is given a concrete example in the article. Melcarek, the aforementioned protagonist physicist, is reported to have solved a problem posted by Colgate-Palmolive, earning $25,000.[6] The article reckons that "paying Colgate-Palmolive's R&D staff to produce the same solution could have cost several times that amount—if they even solved it at all" (Howe 2006). But again, the story is not simply about costs. The reason why Melcarek could come up with the solution, which he admits "was really a very simple solution," while the R&D department at Colgate-Palmolive could not is explained in terms of the structural limitations of the traditional R&D: bureaucratic, hierarchized, stiff, and rigid. While at Colgate-Palmolive's R&D the task of solving a toothpaste tube–related problem was assigned to "test tube guys," the solution Melcarek came up with required "training in physics" (Howe 2006), which he possessed. In other words, for Colgate-Palmolive, crowdsourcing was a matter not just of reducing costs of development but of being able to come up with a smarter solution by tapping into skills and talent they did not have and that in fact they did not know they needed.

Here, too, crowdsourcing is presented as rational and advantageous not only from the point of view of companies and the system as a whole but from that of individuals as well. Melcarek's arrival at the position of a "solver"—as the crowd is referred to on the InnoCentive Web site[7]—is described as the happy culmination of a bumpy professional road. His studies in physics were followed by "a succession of 'unsatisfactory' engineering jobs . . . none of which fully exploited Melcarek's scientific training or his need to tinker"; after all, the article concludes, "not every quick and curious intellect can land a plum research post at a university or privately funded lab" (Howe 2006).

This mismatch between Melcarek's *talent* and a work *position*, in the traditional sense of the term, was accompanied by his reluctance to work in "a 9-to-5 environment." Crowdsourcing is presented as a solution to the limitations put forth by the traditional job market, which consists of *positions* (which are located at definite and stable points in space and time) that are part of a *career path*. These are seen as close-ended and restrictive, in contrast to the open-ended and flexible character of employment afforded by network production, which is specifically adept at exploiting the full potentiality of individuals. From the point of view of his personal work life "for Melcarek, InnoCentive has been a ticket out of this scientific backwater" (Howe 2006). Melcarek's work life is more free and liberated in another sense as well. His work takes place within a physical and temporal environment of leisure. Melcarek's lab is described as a "'weekend crash pad,' a one-bedroom apartment littered with amplifiers, a guitar, electric transducers, two desktop computers, a trumpet, half of a pontoon boat, and enough electric gizmos to stock a RadioShack," and he is doing his work there "on most Saturdays."[8] Having the temporal and spatial freedom that Melcarek enjoys and at the same time taking part in the loop of production is made infinitesimally easier, according to the digital discourse, with network production.

According to the digital discourse, network production brings about a shift in business culture as well. Larry Huston, an executive with Proctor & Gamble, depicts "the current R&D model" as "broken." To overcome this crisis, the current model of tenured positions within R&D departments is replaced by a new employment model, which is flexible, ad hoc, and project-based—in short, networked. The locus of production becomes decreasingly the company, or the organization, and increasingly the network itself, an assemblage of disparate productive nodes that gather ad hoc for a particular project (Castells 1996, 168–72). Hence, from the point of view of companies, network production is seen as a solution to the constraints and limitations set forth by what a *company* is. It is instructive to recall here the etymology of the word. "Company" is a derivative of "companion," from the Latin *com* (with) and *panis* (bread),[9] literally meaning "one who eats bread with another."[10] According to the *Online Etymology Dictionary*, it was first used in the twelfth century to mean "a body of soldiers" and in the sixteenth century came to mean "business association." It is precisely the sense of communality, mutual responsibility, and exclusivity—sedimented in the etymology of the word company—that the new ethos of network production resists, revolutionizes, and in fact wishes to wither away.

Companies therefore increasingly open up to the network of talent, intellect, and knowledge that lies outside of their boundaries and incorporate the new spirit of networks into their operation. Proctor & Gamble's shift to a network model is a case in point. In the past, the company was characterized by an "insular" corporate culture, which won it the nickname "the Kremlin on the Ohio" (Howe 2006). It first adopted the network architecture in-house, realizing that "cross-pollination" between different departments was a good thing for creativity and productivity. Later, it had become even more open, connecting to a much larger network via InnoCentive. This, says Huston, "changed how we define the organization . . . We have 9,000 people on our R&D staff and up to 1.5 million researchers working through our external networks. The line between the two is hard to draw" (Howe 2006).

From the point of view of companies and in terms of production, the line is indeed hard to draw; knowledge and labor pertinent to a particular ad hoc project or product are dispersed over the whole network of production. But from the point of view of workers, the line between themselves and the network of which they are nodes seems, in some respects, to be getting ever thicker and more impenetrable due to the redefinition of "work" and, in effect, the dismantling of this social category.

DISMANTLING "WORK"

In the discourse on network production, workers are redefined as independent freelancers who are compensated solely for their successful accomplishment (i.e., for the end product they sell, rather than for their labor) and who bear all the costs of their work: equipment, training, unproductive time due to illness, health and other types of insurance, and so forth. In other words, they are piece workers.

All these costs are constructed in the discourse on network production as "externalities"—that is, socially necessary costs that are not accounted for in the compensation system (between workers and employers) or in the price system (between sellers and buyers). Instead, these costs are transferred to individual workers—that is, desocialized, or privatized. For example, Melcarek's success story encapsulates—almost by construction—another story that remains latent in the article: that of all the other "solvers" who worked on that particular challenge but could not find a solution or who were too late in finding one. Had they been included in the story, we would have gotten quite a different picture. First, the amount of "brainpower" that had gone to waste (i.e., that was not directed into the accumulation process) would have undermined the universality of the argument that the network architecture marks a greater economic rationality.[11] It would in fact uncover the extent to which a networked mode of production such as InnoCentive is economically more rational *only* from the point of view of companies rather than the system, or society, as a whole.

And second, such depiction would have made clearer the transfer of risks from companies to individuals. This transfer of risks can also be conceived in terms of the privatization of the relations between companies and workers. In network production, companies meet workers at the marketplace to perform a specific task and then part ways. The neologisms that dot the digital discourse succinctly crystallize these new social relations. On the InnoCentive Web site, companies are called "seekers," workers are called "solvers," and work (or the problems to be solved) is referred to as a "challenge." In the mere handful of articles on network production analyzed here, there is almost no mention of the concept of "work" or "labor." The social category and subject formerly known as "worker" has almost completely evaporated in the digital discourse and is instead referred to by a plethora of other labels, which are themselves very revealing. They can be categorized into three fields of meaning—flexibility, democracy, and creativity and passion—all of which succinctly encapsulate the radical transformations of work entailed in network production. Table 5.1 presents a summary of these terms.

While the digital discourse hails the lax, flexible, and ad hoc relations between employers and employees (exemplified in the formula "the strength

Table 5.1. Summary of signifiers for workers in network production in four articles in *Wired* magazine

Category of meaning	Neologism for "workers"	Source
Flexibility	"ad hoc experts"	Howe 2006
	"solvers"	Howe 2006
Democracy	"citizen developers"	Koerner 2006
	"everyday people"	Howe 2006; Kelly 2005
	"users"	Howe 2006; Kelly 2005
	"the people"	Kelly 2005
	"ordinary people"	Kelly 2005
	"the audience"	Kelly 2005
	"regular folks"	Anderson 2006
	"peer[s]"	Anderson 2006
	"fans"	Koerner 2006
	"rowdy rabble"	Anderson 2006
Creativity and passion	"amateurs"[12]	Howe 2006
	"enthusiasts"	Howe 2006
	"dabblers"	Howe 2006
	"passionate, geeky volunteers"	Howe 2006
	"hobbyists"	Howe 2006; Koerner 2006
	"fanatics"	Koerner 2006
	"obsessed fans"	Koerner 2006

of *weak* ties"; Howe 2006; emphasis mine), it suggests that network production actually further ties the relations between the company and worker. Proctor & Gamble's Larry Huston says, "People mistake this [crowdsourcing] for outsourcing, which it most definitely is not . . . Outsourcing is when I hire someone to perform a service and they do it and that's the end of the relationship. That's not much different from the way employment has worked throughout the ages. We're talking about *bringing people in from the outside and involving them in this broadly creative, collaborative process.* That's a whole new paradigm" (Howe 2006; emphasis mine).

The new paradigm, according to Huston, is a tighter, more meaningful collaboration between company and workers. But such collaboration is based on the reconstitution of both company and workers as two independent and autonomous nodes of network production. Paradoxically, the atomization and privatization of work and the worker are presented as preconditions for a new paradigm of production relations, a paradigm that is ultimately more intimate and satisfying. In other words, a process of separation and clear demarcation is needed in order to *then* build healthier relations. The collaboration of these independent nodes in the new capitalism's network production substitutes the symbiotic relations between company and workers in the old capitalism—a paradigm that had proved, according to the digital discourse, to be no longer viable, if not completely dysfunctional.

PASSION AND PARTICIPATION

Another important case study of network production is that of Lego. Its story is told in the article "Geeks in Toyland" (Koerner 2006), which tells of how the Lego company is using network technology in order to harness individuals to help in the development and design of its products. Since this is a cover story, the cover image features a sea of Lego figurines, fading into the horizon. The figurines are all identical; their faces wear an expression of resolute determination (see Figure 5.2). The teaser reads, "The Lego Army wants you: How obsessed fans are helping Lego reinvent the world's coolest toy" (Cover 2006). While the image shows Lego figurines and the title refers to "The Lego Army," both the figurines and the army refer not to something that Lego possesses, but to something Lego can tap into precisely without possessing. The cover art seems to resonate with the fantasy of employers regarding the network: that the network promises a sea of fans, enthusiasts, amateurs, and hobbyists—in short, nonorganized, atomized, and docile units of labor power—eager to be given the privilege to harness their creativity and passion and participate in a business project.

IS THE NEW TECH BOOM FOR REAL? (PAGE 113)

WIRED

THE LEGO ARMY
WANTS YOU
How obsessed fans are helping Lego
reinvent the world's most popular toy

LIFE ON OTHER WORLDS! (We've Got Pictures)
THE PRIUS KILLERS: Inside GM's Chop Shop
One of Our SPY SATELLITES Is Missing

Figure 5.2. *Reserve Army of Enthusiasts.* Cover, *Wired*, February 2006 (reprinted by permissions from *Wired/*© Condé Nast Publications Inc.)

The means by which the crowd is being mobilized for production purposes through network production are varied. Increasingly, these means have become so dominant as to be almost invisible. For example, it has become a commonplace for software companies to release "unfinished" products to the market. This is sometimes done implicitly (such as the case with many Microsoft products, released knowingly with undetected defects, or "bugs") or explicitly (such as the case with "lab" and beta products of Google). In both cases, consumers take an active part (either consciously or unconsciously) in the production process by serving as testers or quality control workers, or even by actively correcting problems and offering improvements. Prosumption in this case entails the transference of part of the production costs on to consumers.

In the case of Lego, the involvement of consumers was prompted at a much earlier stage. When it wanted to upgrade its robot kit Mindstorms, "Lego didn't even have a working prototype. It was way too early for beta testers; Lego needed a Mindstorms User Panel, or MUP, to help with the design" (Koerner 2006). As one of the fans/engineers who was mobilized for the MUP reveals, "I was surprised they were so early in their development, and I think everyone else was, too . . . We realized that our input was going to be a lot more important than we had imagined" (Koerner 2006). In this case, Lego was pushing the envelope in terms of how much of the production process was transferred to consumers.

The digital discourse tells not only of the obvious benefits of network production for producers but also of the newfound opportunities for individuals to collaborate and take a meaningful part in a creative process (as the last quote from one of the MUPers reveals). As the production process becomes networked—that is, more democratic and participatory—people have greater opportunity to take part in shaping the material world they live in: in this case, their toys. As the story continues, it oscillates between the narrative of prosumption as a source of profit and the narrative of prosumption as a new mode of collaboration and individual opportunities. It is instructive to follow how those two narratives are interwoven in the story of Lego's Mindstorms.

To develop Mindstorms, Lego recruited four engineers—who are also "obsessed fans" of Lego—who for eleven months "were de facto Lego employees" with "one key difference between the four panelists and actual Lego staffers: a paycheck" (Koerner 2006). Not only were they not getting paid, they even paid some of the expenses related to their work, such as airfare for meetings in Lego's headquarters in Denmark. The four panelists, however, saw no problem with that: "They [Lego] actually want our opinion?" enthuses one of them. "It doesn't get much better than that." The satisfaction that comes from taking a significant role in a creative process substitutes more traditional forms of compensation; to be part of the productive network is taken to be compensatory enough in the digital discourse.

This enthusiasm translates into consumption practices as well. As the article reports, "such loyalty isn't unusual among the fanboys" of Mindstorms, who also made it Lego's all-time bestselling product (Koerner 2006). The spirit of network production no longer follows the traditional separation between *producers* who labor hard to create a fun toy and *consumers* who play joyously with it; now it is the fun and playfulness of consumers that are mobilized and fed back into the production process. In this move, the toil of tedious, unsatisfactory work clears way for work as fun. And because it is fun—because it is about satisfaction, self-fulfillment, and creativity—traditional pay is cleared out of the way as well.

MOBILIZING NETWORK QUALITIES TO PRODUCTION

From the point of view of Lego, outsourcing the development of Mindstorms had nothing to do with economic gains, according to the article, but rather was a more radical attempt to take advantage of the new and promising qualities of networks. "The boldest part of the Mindstorms overhaul," the article argues, "is Lego's decision to outsource its innovation to a panel of *citizen developers*" (Koerner 2006; emphasis mine). In this language, workers, professionals, and engineers are deemed citizen developers. The use of the word "citizen" distinguishes them from the bureaucratized ranks of Lego employees. Their identity as citizens, rather than the rank and file of the Lego organization, supposedly unfetters in them some quality (Creativity? Joy?) that is deemed missing from Lego employees. For Lego, network production is likened to bringing in a breath of fresh air (with "citizens") and ventilating the stultifying closed ranks of a military-like institution. Like the aforementioned "white lab coat" (Howe 2006), "citizen developers" also connotes the army as the archetypical bureaucratic organization.

Network production, according to the digital discourse, offers a revitalization of a dead-end production process, reigniting it with enthusiasm and fresh creativity. The article even recommends that this production strategy be imported to other Lego products: "Who," it asks rhetorically, "would know better how to improve the company's building systems than the people who spend hundreds of hours preparing for Brickfest [a convention of Lego fans] every year?" (Koerner 2006). Play, creativity, fun, enthusiasm, and leisure—features associated with the consumption of Lego products—are rendered forces of production in the digital discourse. Instead of wasting them, network production allows for their commodification and introduction into the production process of knowledge-intensive components, and in turn increases their value. The move, therefore, is complementary: network production accommodates the "libidinal" energies of workers, and these energies, in turn become new forces of production.

Concurrently, underlying this model is the assumption that since such play is done anyway, it requires no compensation. The rationalization is similar to other cases of nonpaid labor, such as child care and housework, which are perceived to be done by women "anyway," or even "naturally," and are therefore not regarded as demanding monetary incentives and compensations. The article makes this assumption more explicit by theorizing the phenomenon using Eric von Hippel's thesis (put forth in his book *Democratizing Innovation* [2006]), which argues that "the joy of the learning associated with membership in creative communities drives people to generously share their time" (Koerner 2006). Indeed, the relations between consumers and producers brought about by network production are described not with a vocabulary of economic rationality but with terms belonging to the irrational and the erotic; they are based not on interests but on love. "The MUPers," the reporter says, "weren't getting paid. But they were playing a vital role in shaping a product they loved." And one of the MUPers readily admits, "If it had been any company other than Lego, I wouldn't be here" (Koerner 2006).

Prosumption, and more generally network production, therefore, not only reduces the cost of production and taps into the talent of amateurs, or highly skilled individuals, or into the smartness of the network. In addition, it allows the creation of a circular feedback channel—or cybernetics, in Norbert Wiener's terms (1965)—between people's deepest feelings about products on the one hand and the process of their making on the other. Network production—in this case of a Lego product—is able to mobilize a force of production that becomes increasingly significant in the new economy: knowledge. What is more, it is knowledge that seems to be decreasingly that of books and theories (i.e., abstract, universal, rational, and conscious) and increasingly that of the everyday experience (i.e., practical, particular, emotional, concrete, and unconscious). It is knowledge that is made concrete, for example, through playful engagement with a Lego product in one's leisure time.

Whatever it is that "citizen developers" have and Lego employees do not, it is clear that network production is seen as a mechanism of extracting new forces of production from consumers and integrating them into the production process. Forces of production that can in fact only be materialized in the very act of consumption: wants, desires, joy, and so forth. Network production is seen as a mechanism to mobilize this kind of dispersed and intangible knowledge that solely consumers possess. In fact, according to the analysis presented in the article, the incentive for Lego to move to an open network model of production was that its products were beginning to be too cut off from consumer's demands; they began to be too abstract and theoretical.

But what is mobilized from consumers for the process of production is not just their creativity, ideas, and labor power but also their commitment and loyalty as consumers. It is quite ironic that while networks are presumed in the digital discourse to require and predicate an ethic of flexible, ad hoc, noncommitted relations between employers and workers, when it come to the relations between producers and consumers the network is seen as facilitating an ethic of stronger intimacy and tighter relations. As the article suggests, "relying on the MUP is a gamble that Lego hopes will lead not only to a better product but also to a *tighter, more trusting bond* between corporation and customer" (Koerner 2006; emphasis mine). The open source model can help not only in the production process but also in marketing: "Opening the process engenders goodwill and creates a buzz among the zealots, a critical asset for products like Mindstorms that rely on word-of-mouth evangelism." In the spirit of democratization, Søren Lund, the director of Mindstorms, hopes that in the future every Mindstorms customer will "be able to have an effect on how Mindstorms is used and designed." This participation, he further presumes, would have positive effect on sales (Koerner 2006).

According to the digital discourse network production allows for a tighter bond between corporations and customers. The process of how Lego transformed its corporate culture is in fact described precisely as a shift from a company that is too insular and detached from customers' demands to one that reconnects with them through network production in order to make better, more profitable products. The idea of adopting an open source model for the production of Mindstorms came after Lego had its worst year ever in sales. The reason for this downturn is that Lego was coming up with "designs that puzzled rather than entertained [customers]." A Lego executive explains, "We had started to make fire trucks that look like spaceships, building systems that no customer could truly appreciate . . . We had to clean that up" (Koerner 2006). The diagnosis was that something was rotten in the Danish firm; the prognosis called for opening the company up to the outside world, to the network. They decided "to connect customers to the company" (Koerner 2006).

Lego realized Mindstorms needed a more user-friendly programming language. Since "Lego didn't have the expertise to write more intuitive software in house," it outsourced the work. Further down the line of development, when the "Mindstorms team had several mock-ups of new programmable bricks . . . the executives wanted a fresh perspective. And that's when they decided to bring in the Mindstorms users—a community of fanatics who as of 2004 have done far more to add value to Lego's robotics kit than the company itself" (Koerner 2006).

These Mindstorms fans started tinkering with the product and posting their work online. At first, Lego feared of the consequences this might have for its intellectual property rights. Finally, the company "concluded

that limiting creativity was contrary to its mission of encouraging exploration and ingenuity." There was another reason as well: "The hackers
were providing a valuable service" (Koerner 2006). The ethics of "exploration and ingenuity," then, sat well with the notion of network production
as a new source of wealth creation and drove Lego further down the road
of openness to the network. The article sums up the shift in corporate
culture toward the network thus: "Until recently companies were skittish
about customer innovation, fearing that outsiders might leak trade secrets
or that they simply lacked the necessary technical skills. But Lego has
warmed to the power of the open source ethos. It's clear to the Lego execs
that Mindstorms . . . would be a lesser product without the MUPers'
input" (Koerner 2006).

Lego not only tolerated the hacking of Mindstorms but actively
encouraged it, and the company "wrote a 'right to hack' into the Mindstorms software license, giving hobbyists explicit permission to let their
imaginations run wild." As one executive explains, "We came to understand that this is a great way to make the product more exciting. It's a
totally different business paradigm—although they don't get paid for it,
they *enhance the experience* you can have with the basic Mindstorms set"
(Koerner 2006; emphasis mine). Network production then is conceived
in the digital discourse to extract from users a new source of value and
incorporate it into its products. The value produced can be conceived
both as "exchange value" (money) and "use-value" (fun and creativity)
(Marx 1990, 125–31); while much of the exchange value is produced by
consumers, it is transferred almost in its entirety to Lego. Admittedly, the
use-value of the product for consumers is enhanced.

NETWORKS HAVE CENTERS, TOO

While the digital discourse constructs each and every node in the network
as ontologically or qualitatively equivalent to any other node, network
architecture is not necessarily flat and decentralized and can also work to
exacerbate the power of one node at the expense of others. In fact, power
differentials between nodes can be conceived as *relational* and *dialectical* (Bauman 1998; Castells 1996; Sassen 2001). Hence, even thought
the narrative is one of mutuality between corporations and individuals as
simply two nodes on the network of production, a closer reading highlights the asymmetry between the two—an asymmetry that seems to lie
at the heart of network production.

The asymmetry between Lego as a strong node in the network of production and the MUPers as creative and skillful individuals that Lego is
able to harness for its production process is reflected in the photographs

of the main protagonists of the article (Koerner 2006). The photo of the Lego executives is a group photo; they are all copresent at the same time and in the same place.[13] The place is unmistakably a workplace: a large conference table in a large conference room (cutely decorated, naturally, given the nature of Lego's products). In fact, the location can almost definitely be identified as Billund, Denmark, Lego's global headquarters. This corporate photo can be contrasted with the photos of the four MUPers featured in the following page.[14] These are four individual photos taken at different times and in different places. The *mies-en-scène* in all of them is quite evidently not a work environment or an office but the MUPers private residences or private labs. The exact locations cannot be identified and could indeed be anywhere around the globe.

What connects the five separate photos, and what appears in all of them, is in fact the real protagonist of this story—the object, literally and figuratively, of Lego's network production: little Mindstorms.[15] In these visual representations, as in the textual representation throughout the article, the object of desire is in fact quite obscured. Because the object at which the Lego executives fix their gazes in the photograph— regardless of their personal disposition and character—represents desire for profit, growth in sales, and so forth. In contrast, each of the MUPers in the photos is holding an object of quite a different desire: for creativity, self-expression, participation, and innovation. It is precisely the difference between these two different objects of desire that the language of networks—with its emphasis on collaboration and the blurring between different types of nodes—obscures. Or put differently, it is the discrepancy between the two objects of desire that is presumed to evaporate thanks to network production. In the digital discourse, these two very different desires—the emancipatory desire of individuals for dealienation and the system's desire for new sources of profit (i.e., new modes of exploitation)—are presented as complementary and even codependent.

FROM "POWER OF THE PEOPLE" TO "PEOPLE POWER"

This chapter opened with an introduction to the narrative of network production as facilitating individual empowerment and emancipation. Network production, Kevin Kelly maintains, democratizes the means that allow individuals to be more creative and more engaged and turns individuals into participators and collaborators rather than merely competitors in the marketplace. The ramifications of network production on individual empowerment are so revolutionary that Kelly opens his article with a "warning": "beware the power of the people" (Kelly 2005).

This warning receives an interesting twist in an article by *Wired*'s editor in chief, Chris Anderson: Kelly's "power of the people" becomes Anderson's "people power" (Anderson 2006). This shift in emphasis stems from a shift in point of view: while Kelly's article is written primarily from the point of view of the masses—the little people, individuals, nodes in the network—Anderson's is written from the point of view of the "global economy." Anderson's article is the first of six short essays discussing the most prominent "trends driving the global economy," a feature supplementing the special issue of *The Wired 40* for 2006.[16] The most significant trend in the global economy is identified as the collaborative production power of the network: "Blogs, user reviews, photo-sharing—the peer production era has arrived" (Anderson 2006).

The shift in emphasis from "the power of the people" to "people power" is significant in that it displaces the individual from the center of the narrative about network production and recounts it from a more system-oriented view. To put it in sociological terms, it substitutes a macro level structural view for a micro level agency-centered one. Anderson situates the network mode of production within its historical context in a manner that recalls Alvin Toffler's (and to some extent Daniel Bell's) historical materialism—one that puts technology at the heart of social change: "First, steam power replaced muscle power and launched the Industrial Revolution. Then Henry Ford's assembly line, along with advances in steel and plastic, ushered in the Second Industrial Revolution. Next came silicon and the Information Age. Each era was fueled by a faster, cheaper, and more widely available method of production that kicked efficiency to the next level and transformed the world" (Anderson 2006).

For Anderson peer production has become a hallmark of the digital era, and among the most important trends in the new economy, because of its ability to harness new—and hitherto unutilized—sources of power for the production process. Ford's "assembly line" is transformed into network assemblages, and with it changes the whole process of production: "Now we have armies of amateurs, happy to work for free. Call it the Age of Peer Production." And the economy is thus restructured to take advantage of this new power resource. According to Anderson, "From Amazon.com to MySpace to craigslist, the most successful Web companies are building business models based on user-generated content" (2006).

Anderson not only celebrates network production as creating a new source for economic growth but also sees a deeper social transformation brought about by network production. He ties together labor, technology, and democracy in a manner that is particularly revealing: "the tools of production, from blogging to video sharing, are fully democratized, and the engine of growth is the spare cycles, talent, and capacity of regular folks, who are, in aggregate, creating a distributed labor force of unprecedented scale" (Anderson 2006).

The new source of production, the new engine of growth, according to this narrative, is the labor power that hitherto has been left unexploited, that is, has been left outside of the process of capital accumulation. It is comprised of "spare cycles" (i.e., spare time, time that is not yet harnessed for the productive process; time that—contrary to the famous maxim—is not money) and "talent" (i.e., skills, craftsmanship, and knowledge). The new source of production is also predicated on the technological ability to mobilize these "leftovers" of time and talent into the productive (i.e., profitable) process. The more this technological ability improves and the more it is possible to aggregate those forces of production, the more the labor force becomes, in Anderson's words, "distributed." Here the full meaning of *democracy* in Anderson's statement is revealed. The more "fully" "the tools of production . . . are . . . democratized," the more the "labor force" becomes "distributed." Here, democratization is reframed as the decentralization of labor. Democracy, according to this statement, entails more individual autonomy and empowerment and the evaporation of centralized, society-wide, and organized power.

The democratization of production tools creates another mechanism of wealth creation. The new engine of growth is based not only on the conscious mobilization of the spare time and talent of unpaid individuals, "There's also gold in the casual Web droppings we all leave online," Anderson explains (Anderson 2006b). Network technology is well adept in harnessing the noncreative facets of people power as well, such as customer reviews for example, and commodifying them: "Your click trail on Amazon is used to create better recommendations for those who follow. Your query on Google and the pages you find relevant give feedback that fine-tunes the search algorithms. The ads you click don't just boost revenue for Google, they also tell it how much to charge the next advertiser. These companies have found ways to harness the wisdom of the crowd, extracting information that was there all along, just latent and lost" (Anderson 2006). Network technology is particularly adept at tapping into the "wisdom of the crowd"—that market-defined, tautologically self-affirming intellect—and redirecting it into the process of production.

Anderson refers to the nonmonetary nature of network production from the point of view of labor. "Previous industrial ages," Anderson says, "built on the backs of individuals, too, but in those days labor was just that: labor. Workers were paid for their time . . . Today's peer-production machine runs in a mostly nonmonetary economy" (Anderson 2006). He also, as we have seen, stresses the monetary gains for capital from network production. This unequal exchange between "produsers" and companies using network production, are not seen as problematic in the digital discourse. On the contrary: network production signifies a new, healthier relationship between those two ends of the production pole: work and capital. Anderson says, "It's a mistake to equate peer production with anti-capitalism. This isn't amateurs versus professionals; it's each benefiting the other. Companies aren't just exploiting free labor; they're also creating the tools that give voice to millions. And that rowdy rabble isn't replacing the firm; it's providing the energy that drives a new sort of company, one that understands that talent exists outside Hollywood, that credentials matter less than passion, and that each of us has knowledge that's valuable to someone, somewhere" (Anderson 2006). Network production is seen in the digital discourse as a platform for a more reciprocal and satisfying relationship between capital and work—a relationship that is about substituting monetary gains and secure and stable careers with satisfaction and inclusion in the exciting prospects of the network.

CONCLUSION: BETWEEN COOPERATION AND COMPETITION

The chapter has proceeded in two parallel, but also intersecting, lines of analysis. On the one hand, I have presented the central narratives of the digital discourse concerning network production, and have shown the emergence of a new spirit of networks, with its emphasis on the compatibility of the *emancipatory* potentials of network production with its increased *rationality*. On the other hand, I have pursued a more critical analysis that highlights the latent assumptions and conditions of possibility for such assertions. Let me briefly summarize the central narratives of the digital discourse regarding network production and discuss the ideological and sociological significance of the emergence of this new discourse.

According to the digital discourse, the revolutionary nature of the integration of production into network technology is anchored in the creation of more complex and seamless interactions between components of production that had hitherto been separated and dispersed. It is therefore best understood in terms of the eradication of the distinctions between these components: between companies and the network, producers and consumers, producers and users, labor and fun, forces of production and the production process, and so forth. These established industrial demarcations (and more specifically, part and parcel of the Fordist phase of capitalism) are now overturned with the emergence of network production.

The network, according to the digital discourse, becomes the prime locus and axial coordinator of production, replacing the company, the workplace, and the assembly line. It facilitates new productive modes that are predicated on the ability to harness the working power of autonomous and dispersed nodes into the productive process. Moreover, network production is able to mobilize forces of production that have been hitherto unexploited and inaccessible (like free time), elusive (like joy and fun), or highly dispersed (like expertise and knowledge). From the point of view of individuals, these new productive modes allow more people to engage more meaningfully with, and to bring their skills, talents, and passions to bear more fully upon the productive process. The productive process becomes more democratic and collaborative and is geared more fully toward personal fulfillment (of the lifeworld), rather than toward the fulfillment of system's ends. Finally, and crucially, network production makes possible the perfect *fusion* of the needs of personal emancipation with the system's needs of capitalism.

DEALIENATION VERSUS EXPLOITATION

We should understand network production not simply as a description of a new mode of capitalist production brought about by technological networks, but as an analytical framework that at one and the same time also naturalizes and legitimates the new realities, and new constellation of power that this mode entails, specifically, in regards to work. These narratives amount to a discourse where the very notions of work, workers, and by extension the working class, are undermined. "Work" is reconceptualized as an eroticized, playful activity of production and consumption, involving creativity, deep engagement, interactivity, and interpersonal communication. Likewise, the category of "workers" is substituted by a new category of the *prosumers*: an individualized and independent unit of production, consumption, and entrepreneurship.

The discourse on network production legitimizes a shift to a post-Fordist organization of labor and production. The sites where the social organization of work during Fordism (and during much of industrialism) were contained and anchored—the company, the union, the guild, the professional, the cadre—are rendered obsolete with the emergence of the prosumer. At the same time that work becomes more meaningful and humane, allowing greater outlet for personal potential, and harnessing amateurish skills and leisure time into social reproduction, it also becomes more privatized and individualized, shifting more risks from capital to labor, and dismantling the social buffer zone offered during Fordism—a buffer zone that favored social equity and personal security *over* the development of individual potential (Harvey 2005). The discourse on network production legitimates a production process that is at one and the same time more democratic and engaging and undermines the institutional arrangements that made those processes more stable and protective during Fordism.

The discourse on prosumption and prosumers also lends legitimacy to the privatization of work. Workers are reconceptualized in the digital discourse as atomized nodes of entrepreneurship in the network of social production. The seemingly ad hoc, contractual, egalitarian, dehierarchized relations that usually characterize market relations between producers and consumers are transplanted into the relations of production. The discourse on network production is therefore "class-blind"; individuals are reconceptualized as nodes in the network, equal to all other nodes, competing on the shared and leveled playing field of the network. That means that everything that is not translated directly into profit (i.e., everything that is not a contributing component in the process of production) is privatized. Workers become entrepreneurs, selling not their labor power, but the fruits (and only the most ripe of which) of this labor power (on the entrepreneurialization of workers see Huws 2003, 98).

Put in the broader theoretical context of this book, in the digital discourse network production is constructed as a transcendence of the pitfalls of industrial and Fordist production, which entailed a hierarchy between producers and entrepreneurs on the one hand and consumers and workers on the other hand; alienation of the worker from the productive process; suppression of creativity and personal expression; the *massification* of production and consumption, and so forth.

What we see with these new narratives regarding network production follows the key observation offered by Luc Boltanski and Ève Chiapello (2005). The digital discourse offers at one and the same time an *affirmation* and inclusion of the humanist critique of capitalism into the spirit of contemporary techno-capitalism, and a *rejection* and exclusion of its social critique. At the same time that the new spirit of networks promises more engaging roles for individuals in the process of production, it also accepts and naturalizes the individualization, atomization, and privatization of work life, and the liquidation, to use Zygmunt Bauman's (2000) imagery, of protective structures and mechanisms. As it promises more flexibility and creativity, the spirit of networks also accepts and legitimates greater precariousness, instability, and vulnerability.

Moreover, in the digital discourse the humanist demands for decentralization, dehierarchization, debureaucratization, more individual empowerment and satisfaction (i.e., criteria judged by the conditions of the labor process), go hand in hand with the traditional demands of capitalism for productivity (judged by the quality and quantity of the end result of production). Through the axial notion of network production, these two demands are now seen as mutually constitutive. More productivity demands individual freedom, it demands that creativity and personal expression would be brought to bear on the productive process. And at the same time it also demands the atomization of workers, their adaptability and flexibility, and so forth.

Technology plays a central role as a source of social change in the digital discourse, transforming them in a direction that reduces and even eliminates class antagonism, which dominated industrial capitalism, and that was brought under tight political control with the construction of the Fordist social compact (see Chapter 1). As the means of production are decentralized and democratized, the digital discourse presumes, so are social relations in general. Network technology provides a fix that speaks to both *system* and *lifeword* (Habermas 1984). On the one hand, it answers the systemic demands of the economy—rationality, productivity, and growth—by facilitating a mode of production that is more rational and efficient, and musters more creative and productive forces. On the other hand, network technology answers the demands of lifeword, the other half in the promise of modernity for human emancipation and liberation through life activity, through work.

In this respect, the digital discourse represents not disengagement with the traditional critique of capitalism, but a continuation thereof. In the language of Boltanski, the digital discourse shows network technology to be a means that better capitalism in two senses: it gives better responses to the internal demands inherent in capitalism (such as assuring the infinite process of accumulation), while at the same time changes how capitalism works so as to alleviate many of the humanist critiques of capitalism (such as the demand for a less alienating, more satisfying and creative work process).

CHAPTER 6

NETWORK HUMAN

IN THE PREVIOUS THREE CHAPTERS WE HAVE SEEN THE digital discourse's treatment of the integration of network technology and central components of the economic mechanisms of contemporary capitalism: the market, work life, and the process of production. The analysis offered an explanatory framework according to which the digital discourse legitimates the new constellations of power entailed by the rise of the contemporary phase of post-Fordist capitalism. A central concern of these chapters has been humans and how the transformations of market, work, and production transform their lives. After all, they are still the objects and subjects of the digital discourse, a discourse that is concerned primarily (though not exclusively, as we will see in the Chapter 7) with the human condition.

As the material and technological conditions of humans change, so do they, or at least our understanding of what it means to be human. As Gramsci has taught us, the intensified rationalization of production and work that characterized industrial societies at the beginning of the twentieth century also required a new type of human (Gramsci 1971, 297) who would facilitate the emergence of a Fordist mode of accumulation. The new, Fordist "methods of work," Gramsci postulates, "are inseparable from a specific mode of living and of thinking and feeling life" (Gramsci 1971, 302). In the specific context of Fordism, Gramsci writes, "Life in industry demands a general apprenticeship, a process of psycho-physical adaptation to specific conditions of work, nutrition, housing, customs, etc. This is not something 'natural' or innate, but has to be acquired" (Gramsci 1971, 296). A new mode of production and social regulation entails new modes of experiencing life and new perceptions of the individual, its body, and its identity (Lowe 1995; Shilling 2005). Gramsci summarizes the American phenomenon of Fordism as "the biggest collective effort to date to create, with unprecedented speed, and with a consciousness of purpose unmatched in history, a new type of worker and of man" (Gramsci 1971, 302). Thus, with the shift to post-Fordism and new modes of production and work, the current chapter examines the digital discourse on the intersection of network technology and humans.

More specifically, I will ask, What does it mean to be human in a digital civilization? What, according to the digital discourse, is the new conception of humans in a network society? How is the intersection of network technology with humans understood? I will be particularly interested in understanding the ways in which the new network human fits into the networks of market, work, and production—that is, into the post-Fordist situation. The narratives regarding the network human are organized around four facets, or sites, of the human: body, mind, identity, and the unconscious, moving as it were from physical and material facets through the cognitive to self-identification and, ultimately, the unconscious components of the human.

NETWORK BODY

THE BODY SIMULACRUM

Let me begin with an image that appeared in "Electric Word," a long-standing feature in *Wired* capturing the current lingo and discourse of the digital society through short expositions of people, events, technologies, and cultural artifacts. The feature's May 2002 edition presents an artwork by Alba D'Urbano in a short piece entitled "Second Skin" (Moreno 2002). The artwork is a suit tailored from a fabric and printed with an image of the artist's naked body. The photo shows a model wearing a suit of a naked female body (Figure 6.1). Why is this artwork—which has seemingly nothing to do with network technology—featured in *Wired*? What is the meaning of this artwork in the context of the digital society? Why is it an important enough cultural expression—in the eyes of *Wired*—to be counted as an "Electric Word"?

This artwork, I want to suggest, encapsulates a key narrative of the digital discourse concerning the meaning of being human in a digital age. It does that by undermining the modernist conceptions of "the human" as a distinct, autonomous entity demarcated from its environment, quasi-transcendental, unique (or singular) in its capacity for subjectivity. Admittedly subtle and suggestive and through a specific facet of the human—the body—this artwork nevertheless delivers a poignant exposition of the critical stance of the digital discourse concerning the "human" by undercutting humanism's constitutive human/nonhuman distinction, as well as a few other derivative distinctions (subject/object, nature/culture, and human/technology).

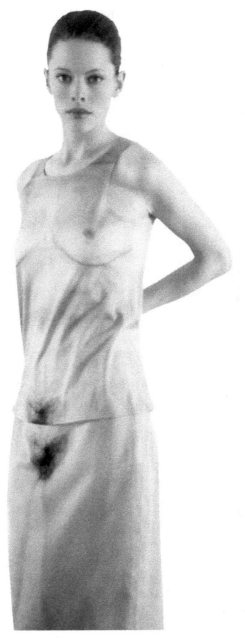

Figure 6.1.   *Inside/out, outside/in.* Artwork by Alba D'Urbano. "Il sarto immortale: Collection" (Detail), 1994–2000. Photo: Gerhilde Skoberne (Frankfurt). Reprinted by permission from the artist.

D'Urbano's artwork renders the human body a suit, which in turn renders the human body. The model therefore covers her naked body with cloths that uncover her naked body. In other words, the artwork does a work of *uncovering* by *covering*, a work of reflexivity, a "movement whereby that which has been used to generate a system is made, through a changed perspective, to become part of the system it generates" (Hayles 1999, 8). In that movement, the rug under the very distinction between covering and uncovering is pulled. As the suit covers the naked body only to reveal the naked body, this artwork also undercuts the distinction between *internal* and *external*. As the text complementing the artwork observes, this work "blur[s] the line between inner and outer" (Moreno 2002). The suit, designed as an external membrane, a "second skin" that covers the first one, offsets its purpose and reveals the internal, first skin.

The artwork undermines another set of distinctions cherished in the humanist discourse of modernity. The body on the suit may look real and natural, but it is really a simulated, produced body. The body, the locus and marker of the human subject, has in this artwork become an object, but a quasi-object at that. However simulated, mediated, and pixelated the body is, it still retains a quasi-subjective quality (we can still claim to have seen *her* naked!). The subject (or a part thereof) can be objectified, while retaining its subjective qualities.

Furthermore, to make the movement between subject and object, between the real and the simulated—between humans and technology— the body should be conceived as comprised of, or reducible and translatable to, *signs*, or information. It is this understanding of the human body and humans in general, as an informational entity that is, as we will see below, key to bridging the gap between information technology and humans and making their conflation plausible.

If the symbolic reproduction of the artist's body as a suit is cautiously titled by *Wired* "Second Skin" (Moreno 2002)—presumably to distinguish this simulated skin from the first, real skin—another artwork featured in the magazine goes further in questioning the distinction between the real and the simulated, the natural and the manufactured, the human body and technology. The artwork is featured in another regular section in *Wired*, entitled "Found: Artifacts from the Future." As the title suggests, "Found" features commissioned artworks that directly engage with a recurring theme of the magazine: technological utopianism (and, rarely, technological dystopianism). The artifacts and their artistic rendering are usually playful and amusing; they represent a fantasy. But a fantasy of what? And what can these fantasies tell us about the ontology of the network human according to the digital discourse? The artwork featured in the June 2002 edition of "Found" is particularity illuminating in the context of the perception of the network human (Bodow 2002).

At the center of this work is a small package designed with the iconography of an existing and popular household object: a box of Band-Aids (Bodow 2002). But unlike the real Band-Aid, which is an external object meant to facilitate the healing of damaged skin, the futuristic product featured in "Found" offers a solution that brings the very distinction between the "artificial" product and the "natural" skin into question. The imaginary product is branded "Quick-Skin" and offers "Self-Grafting Bandages." The pronouncements on the package describe the product as "healing wounds with living tissue"; promise that it "starts replacing damaged skin on contact"; and enthuse about "30 percent faster grafting!" Since both the "first" and "second" skins are made of "living tissue," the very distinction between internal and external, original and imitational, real and simulated, becomes insignificant. As the photo on the package and the samples lying around reveal, while the material may be synthetic in the formal sense (alluded by the maker's name "Plastex Inc.") the validity of the dichotomy between synthetic and organic is brought under scrutiny. Thus, for example, one pronouncement on the package promises "25 nerve receptors/cm^2 for seamless sensory networking" (Bodow 2002). This *seamlessness* implies the complete erasure of difference between the (synthetic) graft and the (organic) skin once Quick-Skin has caught on; if the synthetic material works as if it were a real skin (in terms of look, sensation, and function), then it *is* real. In other words, the definition of the body in the digital discourse becomes pragmatic and instrumental (rather than, for example, transcendental or lyrical).

THE COMMENSURABILITY OF HUMANS AND TECHNOLOGY

Analytically speaking, this is the preliminary task of the discourse on network humans: to insist that network technology and humans are of the same order. The distinction between humans and network technology is blurred and flattened in the digital discourse by placing both humans and technology on the same plane and suggesting that both are paradigmatically and essentially the same. On this plane they might not be quantitatively the same (e.g., a computer processor might be quicker than humans at calculations; a robot may be less competent in visual skills), but qualitatively and ontologically they are perceived to be of the same kind. This narrative of humans and technology sharing the same plane allows for a seamless and frictionless transference between them, a shared discourse of concepts, categories, and imagery where humans (and human qualities and characteristics) can stand for computers and vice versa. By under-mining claims for substantive difference between humans and network technology, two scenarios become possible: the *commensurability* between humans and technology and their *conflation*.

These are not the same as a third, more traditional narrative regard-ing humans and technology: *utilization*. In this narrative humans *use* machines, and there is therefore a clear distinction and hierarchy between humans and machine, subjects and object. The narratives of commen-surability and conflation, on the other hand, transcend the narrative of utility by asserting a deeper similarity between humans and network tech-nology. They presuppose a deeper sense of mutuality, interchangeability, and interactivity. Not only do humans utilize network technology, but also, to a large degree, they are similar to it.

The narrative regarding the commensurable nature of humans and network technology is most evident in the plethora of images that make the transition between technology and humans seem smooth, unprob-lematic, and natural: "seamless," as the expression goes. Let me exemplify this point with three advertisements for digital cameras. It is noteworthy that the three advertisements, for three different companies, appear in the same issue of *Wired*, a testimony to the prevalence of this narrative in the digital discourse and to the homogeneity of its imagery.

While the three advertisements recount essentially the same narrative of the commensurability of humans and information technology, they offer varying degrees of the possibilities of conflation between humans and machines. The first advertisement is dominated by the face of young man pointing a digital camera, ready to shoot a photo. The text, laid out near the eye unoccupied with the camera reads, "careful, this eye may become jealous" (Olympus 1999). While the allusion here is that the camera's visual competence is as good, or better, than the visual aptitude of the human eye, the human-technology distinction still dominates the scene. The serene, warm-colored, smiling face of the man, with its natu-rally curved contours is set in contrast to the cool, metallic, machine-like colors and straight-lined contours of the digital camera.

The second advertisement goes a little further in conflating humans and technology. Here the protagonist of the advertisement is the camera rather than a human. But the human is nevertheless there as an ideal, since the featured camera is purported to be "so smart, it's nearly *human*" (Kodak 1999). The human—or rather a human quality—is "captured" by the camera; its eye peeks behind the camera's lens. This time, the warm colors are shared by both the camera and the human. Like the previous advertisement, the camera is anthropomorphized and is assumed to have human qualities. The camera's superior ability to perceive reality visually is equated with that of a seeing human. (And the caution appearing in the previous advertisement is affirmed: the eye might indeed get jealous, as the camera's visual ability does rival it.)

The third advertisement takes the immersion of humans and technology one step further. In contrast to the previous advertisements, where the camera is featured front and center, in this advertisement the camera seems to be missing from the scene altogether. Instead, a woman's face overflows the page, dominating the scene. Her face is flooded with a cool, bright blue light, which diminishes much of her "humanness" and gives her a semihuman (and perhaps semi–science fictional) look. A closer look into her eyes—the most expressive and revealing elements in the face, the "gateways to the soul"—reveals that they are actually "made of" digital cameras. The slogan for this advertisement reads, "Change the way you see the world" (Sony 1999a), which imbue the digital camera with the ability of "seeing."

HUMANIZING TECHNOLOGY, TECHNOLOGIZING HUMANS

By focusing on the technological, visual skill of humans, what these advertisements are doing, to a greater or lesser degree, is to flatten out and diminish the distinctions between humans and technology. Digital cameras are perceived as seeing organs to the same extent that humans are seen as organic cameras. The very distinction between organic and synthetic, nature and culture, created and manufactured—in short, between instruments and technology on the one hand and humans on the other hand—is diminished by focusing on the *instrumental* and *technical* aspect of the human body.

The digital discourse positions humans and technology on the same plane, hence lubricating, as it were, the channels of transferability and commensurability between them. An advertisement for Sony further exemplifies this point. The advertisement features four sequential full-page images. The first features a blue-lit woman's face with camera eyes, and the text reads "see it." The next two images follow the visual style of the first. In the second image components from an audio device are grafted onto a woman's face, and the text reads "hear it." And in the third, the back of a human's head is adorned by a slot into which a memory card is inserted, and the text reads "think it." The fourth page resolves the mystery of the preceding hybrids by exclaiming "stick it," featuring various information technology gadgets (Sony 1999b). This advertisement admittedly sells memory cards and electronic equipment, but it also propagates an idea: that the richness of humans' senses, experiences, and intellect can be easily transported and recreated with information technology;

that visual and auditory sensations, as well as thought, can be uploaded on, and downloaded from, humans and information technology.

The body in the digital discourse is perceived as an instrument, the material and functional levels of which can be increasingly rationalized and denaturalized, as something that can increasingly be artificially manufactured, or virtually represented, rather than naturally created, in a word, as technology. As humans—their bodies, their functional abilities, and their characteristics—come to be seen as objects and technologies, the prospect of a modular construction and reconstruction of machine-like humans and human-like machines comes to the fore in the digital discourse. One scenario brought up by the objectification of the body is the manufacturing of technologies (organs and body parts) to be incorporated into humans, as the subtitle of "Regrow your own" (Kahn 2003) cheerfully proclaims, "Broken heart? No problem. New liver? Coming right up." (The image on the cover of the September 2000 issue of *Wired* tells of another fantasy of meshing humans with machines; see Cover 2000). Clearly this smooth transferability is seen as empowering and liberating.

The other side of the coin of the objectification of the body is the prospect of producing human-like robots, "Humanoids" (Capps 2004), or "Robosapiens" (D'Aluisio 2000). An exemplary article in this vein offers a catalogue of technological projects currently underway or in planning stages that "might one day add up to an android just like the rest of us" (Capps 2004). The six-page catalogue presents a dissection of this future robot into muscles, hands, expression, lips, skin, eyes, ears, nose, stomach, and legs.

It does not make much difference that these are futuristic presentations or that the digital discourse also asserts that the human body is still superior to robots on many counts. What is crucial here is that as we are presented with human-like machines, we are also encouraged to think about humans as machines. That is, we are set within a dual framework of instrumentality and commensurability. Instrumentality pertains to seeing the body as a technology, a functional tool, and an instrument; commensurability pertains to the ability to see humans and technology as interchangeable and conflatable. Robots are therefore central in the digital discourse not only because of their potential service to humans but also, more importantly, because robots—as a discursive category—play a central role in redefining the identity of humans. As such, they dot the digital discourse and are often reported on in *Wired* in the context of interaction and conflation rather than utilization. This is evident, for example, in the words of an MIT scientist in the introduction to "The Humanoid Race" (Capps 2004), who predicts that "within five years, robots will cross a critical threshold, becoming partners rather than tools—in other words, we'll have friends, not appliances" (Capps 2004).

And on emotional and conceptual levels, this threshold seems to have already been crossed; in the digital discourse robots are already recognized as humans. "The Humanoid Race" (Capps 2004) features nineteen different robots, with each receiving a *personal* representation (i.e., not as a type, but as a particular): each has a distinctive look and a distinctive name (such as Jerry and Asimo). But they are all—after all—robots; part of "the family of post-man," to paraphrase the title of the famous 1950s MoMA exhibition (Steichen 2002).

To the degree that humans, and specifically in this case their bodies, are seen as technologies, they are also instrumentalized and commodified. They become things, objects that can be evaluated quantitatively (in terms of price and performance) and, in turn, be improved, enhanced, altered, and purchased. As instruments, human beings also become a means; a clear break from an Enlightenment humanist notion of humans as carrying substantive and unqualified worth as subjects and as ends in and of themselves.

One of the most theoretical articulations of this posthuman stance is put forth by Donna Haraway, whose thought is featured in "You Are Cyborg" (Kunzru 1997). Haraway's basic thesis, according to the article, is that "the realities of modern life happen to include a relationship between people and technology so intimate that it's no longer possible to tell where we end and machines begin." She prefers, therefore, to speak of cyborgs rather than humans. For Haraway, the technologization of the body entails not only the physical internalization of technology into the body but also the "idea of the body as high-performance machine"—that is, the instrumentalization of the body.

Haraway sees no moral dilemma concerning the "cyborgization" of humans. Referring to the use of performance-enhancing drugs in sports, "Haraway could hardly see what the fuss was about. Drugs or no drugs, the training and technology make every Olympian a node in an international technoculture network just as 'artificial' as sprinter Ben Johnson at his steroid peak" (Kunzru 1997). Haraway refuses to differentiate culture (or nurture) from nature, since she refuses to see the body as natural in the first place. "Haraway's world is one of tangled networks—part human, part machine; complex hybrids of meat and metal that relegate old-fashioned concepts like *natural* and *artificial* to the archives" (Kunzru 1997).

Not only is the cyborgization of humans untroubling, but also, for Haraway, this reconceptualization of humans as networks is liberating, particularly for women. The cyborg is a "symbol of feminist liberation," since it undercuts the naturalistic ideology that legitimizes women's inferior social position. "Women for generations were told that they were 'naturally' weak, submissive, overemotional" and so forth. "If all these things are natural, they're unchangeable . . . On the other hand, if women (and men) aren't natural but are constructed, like a cyborg, then, given the right tools, we can all be *reconstructed*" (Kunzru 1997).

This, according to the article, is also the main thrust of cyberfeminism, which promotes the idea that "in conjunction with technology, it's possible to construct your identity, your sexuality, even your gender, just as you please" (Kunzru 1997). One of the most obvious and intrusive realm of technologizing of the human has been genetic engineering, where "we are becoming the objects of our own technological processes" (Simpson 1997) and where "man create[s] man in his own image" (Rohm 2004).

Haraway, and cyberfeminism in general, see the liberating potential of the discourse on the network human in rendering humans malleable and flexible, readily available for continuous, multiple, and ad hoc deconstructions and reconstructions according to some nonessentialist and *extranatural* code. In the absence of any essentialist natural code, the theoretical question then becomes, What is the external code for the construction of the human? As we will see below, it is the characteristics of the network—as constructed in the digital discourse—that come to dominate the understanding of humans.

Network Mind

What the digital discourse suggests is that the intensity and profound level of interaction between humans and technology may imply that we can no longer speak of *inter*action, which assumes the independence of two entities and assigns technologies the role of objects and humans the role of subjects. Instead, the digital discourse suggests, we are witnessing the merging of humans and network technology into a singular entity; or to turn this proposition upside down, we see the emergence of a new subjectivity comprised of both humans and technology.[1] This idea receives an eloquent representation in an artwork featured in *Wired*. The black and white photograph by artist Matthew Pillsbury depicts a quotidian moment in the life of the network human: a dining table inside an apartment, with three laptop computers on top (Brown 2004; Figure 6.2). Both the title of the artwork and the complementary text initially suggest a straightforward depiction of a trivial scene. The work is entitled "Penelope Umbrico and Lila and Rebecca Cohn," and the text suggests that it "captures a mother and her daughters as they check email, send instant messages, and tend to their virtual pets over the course of a half hour" (Brown 2004).

But something weird goes on in this otherwise banal representation, expressed by two visual dimensions that seem to dominate the scene: a dimension of clarity and blurriness, and a dimension of lighting and shading. As for the first dimension, the clearest objects in the photo are those that are the most immobile, the "hardware," as it were: the walls, the hanging lamp, the plants, and the three laptop computers. In contrast to these, the computer users look blurred, fuzzy, and almost nonexistent. As for the second dimension, while the apartment is generally dark, the two centers of extreme brightness (or "overexposure") in the photo are the three computer screens, and the three heads of the people using the computers.

The combination of these two visual dimensions creates an image of great poignancy: the banal, calming scenario of this home is a background to a much more disquieting drama that takes place in its midst—that is, of humans merging with their computers. This is obviously not a physical

Figure 6.2.   "Penelope Umbrico and Lila and Rebecca Cohn." Artwork by Matthew Pillsbury. (Copyright Matthew Pillsbury/Courtesy Bonni Benrubi Gallery, New York City)

merging but a merging of software: human consciousness merges with cyberspace. The different physical shape of humans and computers is only a cover-up for a convergence of similarities—their consciousness. As the complementing text asserts, in this artwork, "subjects don't just stare at their computer screens, they vanish into them. Pillsbury's work . . . demonstrates how humanity and technology are intertwined" (Brown 2004). Thus, humans vanish into their computers, creating a seamless flow of consciousness between the two and, in fact, constituting each as nodes in a network. As this happens, the independent subjectivity of humans is undermined; their consciousness and thoughts are fundamentally intertwined with the machines they use. This not only makes humans quasi-subjects but also endows machines with a quasi-object status as well (Latour 2005).

It is worthwhile at this point to pause and emphasize the full significance of this likening of humans to machines. In the digital discourse, this likening is merely a prelude to a much more profound claim: that humans and network technology are commensurable to such a degree that they can merge into unity. This likening should not be underplayed as merely metaphorical; rather, it is substantive and, hence, instructive of the horizons opened up by such a discourse. Consider a more common metaphor: "time is money." To treat this maxim—immortalized in eighteenth-century America by Benjamin Franklin—strictly as a metaphor implies that time has similar characteristics to money in that both are quantifiable, malleable to manipulation (such as loss or accumulation), dear, finite, and so forth. As a metaphor, "time is money" imbues time with a very particular set of meanings (associated with capitalism) rather than alternative sets of meanings.

But that maxim carries an even heavier cultural significance. Once time is shown to be *commensurable* with money, once proper practices are introduced by which time can *really* be translated into money, then this can no longer be treated as a mere literary metaphor; rather, it becomes an instrumental and practical equation. Thus, for example, time becomes money (i.e., it can be translated into money) in the most pragmatic and instrumental sense under the specific conditions of capitalism, which, as both Marx and Weber pointed out, is conditioned by the emergence of wage-labor. As Marx has shown, time becomes particularly crucial under capitalism, where value that can be diverted to capital accumulation rather than reproduction comes from the difference between the total hours of work and the hours needed to create the value required for their own reproduction—that is, their wages. Marx calls this difference surplus value. This "trick"—translating the laborer's (work)time into the capitalist's capital—is only possible with the commodification of labor—that is, with the creation of wage laborers who are compensated for their *time* and not for the market value they actually produce.

Hence, in the historical context of its coinage, Franklin's maxim is not merely a metaphor regarding the essence of time but an ideology, at one and the same time revealing and concealing the realities of capital accumulation under capitalism and its connection with time. "Time is money," then, is rendered a strong metaphor when these very different entities are shown in practice to be commensurable. To tie this back to our discussion here, the discursive intensity and potency of a metaphor increases (and becomes practical) when some form of commensurability is pointed out and some channel by which the commensurability of humans and network technology can be materialized.

How is the commensurability between humans and network technology possible according to the digital discourse? The answer, as we have already begun to see, is that both humans and network technology share the common denomination of being essentially informational machines. Thus, humans and network technology are commensurable through the media of information. This commensurability through the medium of information (or more precisely the medium of binarism or digitalism) allows the connection of humans to technology. Thus, for example, the digital discourse renders the brain a "hardware" on which "software," or "programs," are run. Thus, whether the hardware is made of neurons or wires, and whether the program is run by a string of binary neural activity or electric signals, becomes insignificant. Unsurprisingly most of the stories in *Wired* that feature the narrative of commensurability focus on the brain as a site of interconnection between humans and machines: the brain is referred to as a "humans-machine interface" (Freund 2004), and the process of interaction is referred to as "brain-computer interface" (Martin 2005) and "neural interfacing" (Branwyn 1993).

DISEMBODIED CONSCIOUSNESS

But in the digital discourse, the brain is often just an entry point (and a euphemism) for a more substantial discussion concerning the core elements of being a network human: thought, mind, consciousness, and unconsciousness. "Mind Control" (Martin 2005), a story about a cyborg, is instructive in uncovering the digital discourse's underlying assumptions regarding the human mind and its compatibility with network technology. The story focuses on Matthew Nagle, a quadriplegic, paralyzed from the neck down, and the first person to use BrainGate, a system that allows him to manipulate information on computers solely with thought. The system is comprised of an array of microelectrodes that are implanted in his brain and register its neural activity. A computer processor then recognizes the patterns associated with arm motions and translates them into signals on a computer screen. Simply put, the system bypasses and substitutes body movements hitherto needed to execute informational commands from brain to computer (such as moving a computer mouse with one's hand), rendering human-computer interaction more direct and disembodied.

To make the leap between human thoughts and computer activity plausible, "thoughts" are equated in the digital discourse with "neural activity," which in turn, is equated with "digitalism." Thus, for example, at the end of a trial session with Nagle, the experimenter renders the sounds of the neural activity audible on a loudspeaker. The author then comments, "A loud burst of static fills the room—the music of Nagle's cranial sphere. This is raw analog signal, Nagle's neurons chattering. *We are listening to a human being's thoughts*" (Martin 2005, emphasis mine). The brain's neural signals might be described here as analog, but it is precisely because they can also be thought of as *equivalent* to the computer's digital signals that the two can be unproblematically juxtaposed. That is the only way to connect these neural signals (or "thoughts") so easily with information technology, which is also based on the neural-like activity of binary units. This, according to the digital discourse, is the conceptual and practical level where humans interface with information technology, and this is how the corollary conclusion of the author can be asserted: in the final analysis, this neural activity *is* human thought. To the degree that thought and consciousness are reduced to neural activity, they can also be understood in terms of information technology.

The digital discourse offers a more general theory of consciousness based on a distinction between the content of thought and its media of production, transmission, and reception—that is, between software and hardware. This is very much in line with the foundations of information theory laid out by Claude Shannon and Warren Weaver in 1949, a theory that largely underscored information technology and paved the road to its development (Shannon and Weaver 1948). According to this theory, human brains can be likened to computers. This narrative of the brain-as-computer (and vice versa) is well known and has indeed become a staple of public discourse and popular culture. Underlying such theory of consciousness—based on the comparison of brains of computers—are two forceful assumptions about the human mind and human consciousness. Not only does it assume a *disjuncture* between, and *independence* of, media and content, brain and thoughts, hardware and software, instrumentality and substance, means and ends, but also a *hierarchy* between them. Both of these assumptions are central to the new conception of the network human.

These two assumptions underlay, for example, the story about Mattew Nagle in *Wired*. The author takes the "hardware"—the brain—as a starting point to ponder about consciousness. "Roughly the size of deflated volleyball," he says, "your brain weighs about 3 pounds. Its 100 billion neurons communicate via minute electrochemical impulses" (Martin 2005). This hardware, according to the author, this network of neurons, *then* produces a variety of "software," or thoughts: "From this ceaseless hubbub," he says, "arose *Ode to Joy*, thermonuclear weapons, and *Dumb and Dumber*" (Martin 2005). The article presumes thoughts and consciousness to be by-products of the mechanics (or rather electronics) of the brain and, as such, to be analytically, temporally, and structurally secondary.

They are secondary in another sense as well: while the ("hardware") brain is universal and value-neutral (an instrumental means; "software") thoughts are contingent and subject to judgment. The same bundle of neurons, with the same binary operation, is shared by all humans; it is presocial and precultural. It is in this sense that it is *hard*ware: stable, robust, and unchanging. *Soft*ware, however, is more ephemeral, contingent, fleeting (a thought might "spring to mind" or "slip one's mind"), and diverse; it is cultural, infinite (one can always come up with new thoughts, new content), and subject to tests (in juxtaposing *Ode to Joy* and *Dumb and Dumber* the author of the article clearly exercises here his Kantian critique of judgment). The corrolary is that the *brain*, the computer-like brain, is the unchanging, ahistorical essence of the human mind and human consciousness. It is this dichotomy, and this hierarchy between brain and mind, neurons and consciousness, that allows the author to suggest that the "the consciousness problem"; that is, the question of how "electricity and meat make a mind," which has been puzzling scientists for almost a century, has reached at least a partial resolution with scientists now being able to "listen in on—or alter—brain waves" (Martin 2005).

HUMANS ARE COMPUTERS, TOO

In the digital discourse, the unchanging hardware ("brain")—not the contingent software ("thoughts")—is seen as the locus of human essence. This comes out very clearly in a story in *Wired* reporting on the development of a new type of lie detector (Silberman 2006). In this story, the disjuncture and hierarchy between brain and consciousness and the ability of full commensurability between the brain and computers (as two types of binary hardware) reaches an ironic moment where the brain unwittingly colludes with the computer, bypassing human consciousness (and human agency) altogether. The novelty of the new lie detector lies in the ability of the new technology to go "deeper" into the brain to get, as it were, to the "heart" of where lying *really* takes place.

This new method is compared with the old, analogue lie detector—the polygraph. The latter measured surface, bodily phenomena such as heart rate and rising blood pressure. These measures change with increased levels of stress and, in turn, indicate whether the person is lying. The polygraph assumed a link between stress and lying. The problem with this assumption, the author says, is that "Sociopaths who don't feel guilt and people who learn to inhibit their reactions to stress can slip through a polygrapher's net." The new lie detection technology, the fMRI, "should be tougher to outwit because it detects something much harder to suppress: neurological evidence of the decision to lie" (Silberman 2006).

The new technology goes to the "source code" of lying, bypassing the conscious human and undermining his agency. The story is accompanied by brain images produced by the fMRI, which supposedly distinguish a truth teller from a liar (Silberman 2006). The article interprets them thus: "When someone is telling the truth, the areas of the brain shown here in green become active. If he is lying, the parts of the brain shown in red display even more activity" (Silberman 2006). By human-machine commensurability and by establishing a direct link between the human brain and a computer, we can eliminate the soft, superficial, weak human epiphenomena and go directly to the hard, essential elements of humanness. Humans might sell a lie as truth to another person; brains, however, cannot pull the same trick on computers. The partnership of brain and fMRI gets rid of another human agent, the polygraph reader. In the new method, reading is done by the computer in order "to eliminate one major source of polygraph error—the subjectivity of the human examiner" (Silberman 2006). This method effectively establishes computer-to-brain communication, eliminating the subjectivity and agency of humans.

Since the essence of humans lies in their brains (i.e., in them being ultimately informational and computational devices), and since, as such, they can be incorporated into computers, information technology is understood in the digital discourse to partake in the construction of human selves. This point can again be exemplified with the story of the aforementioned Matt Nagle. As a cyborg, an entity conflating a human with information technology, Nagle, the article insists, is more than simply "part biological, and part silicon, platinum, and titanium" (Martin 2005). Just as information technology is incorporated within him, part of him is externalized to a computer.

The identity of the cyborg is presented in the digital discourse as a matter of construction not unlike any other identity; it is not a natural and direct derivative of the hybridization with technology, just as sexual or gender identities are not direct and unproblematic derivatives of one's genitalia. To become a cyborg, Nagle had to learn how to use his new computer "organs" and to adopt new ways of thinking, or a new *mind*set, inextricably intertwined with information technology. Hence, after having the operation that meshed together flesh and metal in his body, he went through a series of "encounters" with the computer. At first, the researcher "asked Nagle to think left, right, then relax" (Martin 2005). The monitors reveled that the computer clearly picked up the neural activity. That was Nagle learning to use BrainGate.

But these encounters were as much about training the computer as they were about training Nagle; the computer had to be acquainted with Nagle. The reporter describes the following: "I watched [as the researcher] takes Nagle through a typical training session. He tracked Nagle's mental activity on two large monitors, one of which displayed a graph of red and green spiking lines. Each spike represented the firing of clusters of neurons. As Nagle performed specific actions in his mind's eye . . . the electrodes picked up the patterns of nearby neuron groups. Then BrainGate amplified and recorded the corresponding electrical activity. Over dozens of trials the computer built a filter that associated specific neural patterns with certain movements . . . The machine was memorizing Nagle's characteristic neural firing patterns" (Martin 2005). This mutual training and learning cannot be underestimated. Since as the machine and the human form mutual recognition and become acquainted with each other they become, according to the digital discourse, indistinguishable, they merge into a singular, posthuman entity, a cyborg.

DISTRIBUTED SELF

For the cyborg Matt Nagle, *being* entails a close interaction with information technology. The author captures this narrative succinctly after a short conversation with Nagle: "What are you thinking about when you move the cursor? I asked. 'For a while I was thinking about moving the mouse with my hand,' Nagle replied. 'Now I just imagine moving the cursor from place to place.' In other words, *Nagle's brain has assimilated the system. The cursor is as much a part of his self as his arms and legs were*" (Martin 2005, emphasis mine). Together with the computer, Nagle forms a network, and so he, too, becomes networked and distributed among various computational devises: his brain, BrainGate, and so forth. That is true both in functional terms and in terms of his sense of self and identity.

The digital discourse on the network human mimics some of the postmodern sensibilities regarding identity, particularly ideas regarding the constructed and nonessentialized character of identity. But paradoxically, perhaps, the digital discourse also carries strong essentialist assumptions that are closely linked with network technology. According to the digital discourse, network technology resonates with the essence of humanness. Only this essence is now presumed to reside in the human hardware (i.e., the brain) rather than the software (i.e., personality, character, and so forth). This anchoring of human ontology in the brain is manifested, for example, in a story about skills of the informational elite, "Revenge of the Right Brain" (Pink 2005).

This article is adapted from a book titled *A Whole New Mind: Moving from the Information Age to the Conceptual Age* (Pink 2006). The author argues that, with the further advent of the information revolution, workers in the West will increasingly have to rely on conceptual skills associated with the right hemisphere of the brain rather than those associated with the left hemisphere. The left hemisphere, which features "computer-like logic, speed, and precision" and "handles sequence, literalness, and analysis" dominated the skills of knowledge workers of the last three decades. But its importance in the West is now undermined by outsourcing ("foreigners can do left-brain work cheaper") and information technology ("technologies are proving they can outperform human left brains"). To find new sources of wealth creation, knowledge workers in the West have to switch on their right hemisphere, which "takes care of context, emotional expression, and synthesis" (Pink 2005). What is distinctive in this analysis is the *dissociation* of human consciousness into two distinct entities and the *association* of one to computers. This dissociation is even more evident in the illustration that accompanies the article (Pink 2006). The illustration can in fact be read as a most precise and direct expression of the spirit of networks. The left brain represents the critical stance of the spirit of network against the human condition during Fordism: bureaucracy, uniformity, impersonality, boredom (represented with a gray-blue monochrome), lack of personality and creativity (all workers look exactly the same), lack of flexibility (represented by straight lines, and the closed and suffocating office space). The right brain represents the new spirit that responds to the humanist critique: personal expression, liberation, creativity, joy, the blurring of leisure and work, exploration, and eroticization. Moreover, the historical shift in capitalism is represented here quite directly as the escape of people form the left brain to the right.

To the extent that *network* humans start to understand their self as such, they also begin to understand themselves in terms of network technology. Thus their personal qualities and characteristics take the form of the technology. This is evident, for example, in the case of connectionism as a prime and valued quality of the network human. Much of the power and prestige of the network human stems from her skills of what Boltanski and Chiapello term connexionism (2005, 362). Esther Dyson—described in *Wired* as "the most powerful woman in computing," and "The intellectual of computer analysts"—is celebrated in particular for her networking skills. As she acknowledges, "Some people like to do the work. I just like putting people near each other" (Borsook 1993).

The network human does not produce as much as he or she brings together and catalyzes the creation of assemblages of disparate nodes in the network. Stewart Brand, for example, publisher of the *Whole Earth* catalogue and an iconic member of the digerati (see, for example, Turner 2006) is described as "a starter of things, an intellectual pies piper with a knack for bringing together people from a wide swath of disciplines" (Hafner 1997). These new types of people, able to transcend disciplinary, professional, and bureaucratic boundaries and bring together disparate nodes, are referred to in the digital discourse as "The Connectors" (Howe 2003): "Meet the hypernetworked nodes who secretly run the world . . . In a computer network, a node performs the crucial task of data routing, playing digital matchmaker to packets of information. In a social network, a node is the person whose PDA runneth over with people they met once on an airplane" (Howe 2003). The network structure creates and is dependent on a new type of social actor whose prime practice is to make connections: the network human.

NETWORK IDENTITY

The type of cyborg represented by Nagle—a nearly science-fictional character of flesh and bolts—is only a borderline case of a much more universal phenomenon shared by an increasing number of humans. To recall Haraway's contention, we are all cyborgs, and we are increasingly so with the widespread dissemination of network technology. According to the digital discourse, in the digital era humans cannot be understood independent of network technology. Information technology is no longer a tool to be utilized but a component of our innermost realms; it also becomes a central axis of human identity. It is here, in the realm of identity, reflexivity, and self-understanding, that commensurability of humans and network technology is most dramatically revealed to be understood not only in instrumental and pragmatic terms but also in quasi-essentialist terms. Humans are understood in the digital discourse to be not simply like network technology but to be themselves networked entities.

A crystallized and systematic articulation of the conflation of network technology and human identity can be found in two articles focusing on the work of Sherry Turkle. The article "Sex, Lies and Avatars" (McCorduck 1996) asserts the compatibility of network technology with the essence of human identity. Both, it contends, are postmodern. In contrast to modernism, which is concerned with form and essences and "assumes that beneath any surface exists a timeless and placeless truth," postmodernism, according to the article, presumes that "no unitary truth resides anywhere." For postmodernists, "there is only local knowledge, contingent and provisional . . . The surface is what matters, to be explored by navigation, not by opening up the hood and peering inside . . . Postmodernism celebrates this time, this place; and it celebrates *adaptability, contingency, diversity, flexibility, sophistication, and relationships*—with the self and the community" (McCorduck 1996, emphasis mine).

That this language, capturing the spirit of postmodernism, bears such an uncanny resemblance to the discourse on network technology is hardly incidental (a resemblance that does not evade the author of the article, who comments that this list of characteristics "sounds like the World Wide Web? Like your favorite electronic game?" [McCorduck 1996]). Turkle's argument regarding the fit of network technology to the spirit of the time underlies the article. Modernism, Turkle says, is "the spirit of the Tofflerian Second Wave"—that is, industrialism. Postmodernism, in contrast, is "the spirit of the Third Wave," a wave, according to Toffler, brought about and defined by information technology.

Here the notion of a break within the history of information technology (discussed in Chapter 5) is reiterated: information technology sprouted from the industrial, modernist world and was reproducing its centralized, hierarchized, system-oriented characteristics, but it developed and was indeed revolutionized in such a way as to enhance personal freedom and unleash human potentialities (see Dean 2002, 79–113). As mainframes—that is, the large, centralized machines of the early days of computing—the article upholds, computers were modernist, but with the introduction of the personal computers, "computing slipped into postmodernism." Now, with the advent of the Internet, "computing continues its postmodern odyssey . . . to the most dramatic extreme: the creation of online communities containing online personae. With its screen surfaces, its learning by doing instead of learning the rules first, its hypertext . . . computing now is as postmodernist as it gets. It's characterized, as Turkle puts it, by 'the precedence of surface over depth, of simulation over the real, of play over seriousness' . . . Computing also offers pluralism, different things for different people" (McCorduck 1996).

How does this bring to bear on human identity? In the weak version, the article suggests that humans now form their identity through information technology. This is in line with the technology-shape-society argument (see Chapter 1). But one can also read a stronger argument, according to which human identity is *always already* informational and networked and has therefore found its match with information technology, since both share a common *essence*. This argument assumes an ahistorical, essentialist human nature. The author reflects on the compatibility of human identity and information technology through Turkle's personal biography, thus exemplifying and affirming it: "In *Life on the Screen*, she [Turkle] describes a shy, English-speaking Sherry, who, to her astonishment, is replaced in Paris by a much more assertive and self confident French-speaking Sherry. This self-transformation was happening inside language . . . And it was language—slippery, ambiguous, elusive, potent—that would be the construction materials of the Internet, too" (McCorduck 1996). According to Turkle, information technology, specifically cyberspace, becomes the site through which human identity is both constructed and uncovered. Postmodern theory becomes the discursive glue that ties these two moments together. In multiple passages the fit among postmodernist philosophy, human identity, and network technology is asserted:

> Turkle's study of Lacan . . . was preparing her for a future she couldn't anticipate: a future represented by computing. Computing would offer her endless moments of sweet epiphany, when theories that had seemed right but abstract were suddenly right and manifest . . . Constructing the self with language and the notion of permeable boundaries? There it was on the screen. You could almost substitute computing for the terms of Lacan's manifesto . . . computing is what brings postmodernism down to earth . . . the computer has transformed the knotty difficulties of postmodern theory into the trivially obvious . . . [Turkle's] findings—that identity isn't fixed but variable and fluid—are very postmodern. And those findings are based on what's actually happening [in cyberspace], not on philosophical speculations . . . What in other contexts has seemed like the gibberish of postmodernism—decentering (oh, you mean multiple users), intertextuality (oh, hypertext), fragmentation (oh, me in the Parenting conference, me in the Eros conference), blurring (oh, object-oriented languages)—is rendered clear at last . . . Postmodernism. Right in front of you. Your laptop embodies new ways of thinking, carries you to them (or them to you), and opens you to them. People use the Internet as "a significant social laboratory for experimenting with the constructions and reconstructions of self that characterize postmodern life". (McCorduck 1996)

THE CO-EVOLUTION OF HUMANS AND COMPUTERS

Turkle, according to the article, argues that as we interact with network technology we become more like it. Or rather, *who* humans are becomes entangled with *what* network technology is. This is the thrust of another article in *Wired* written by Turkle herself. Turkle's main argument is that "computers are not just changing our lives but changing our selves" (Turkle 1996).[2] She begins her article with an answer to the question posed in the title—"Who Am We?"—that asserts her multiplicity: "There are many Sherry Turkle." She explains how this multiplicity is reflected, facilitated, and augmented by network technology. Here, for example, she makes this point with reference to the Windows operating system: "Windows have become a powerful metaphor for thinking about the self as a multiple, distributed system" (Turkle 1996). Later she adds, "Those boxed-off areas on the screen . . . allow us to cycle through cyberspace and real life, over and over. Windows allow us to be in several contexts at the same time—in a MUD, in a word-processing program, in a chat room, in e-mail . . . The self is no longer simply playing different roles in different settings at different times. The life practice of windows is that of a *decentered self that exists in many worlds*" (Turkle 1996, emphasis mine) Turkle quotes one of her research subjects saying, "Now real life itself may be . . . just one more window" (Turkle 1996). Turkle, then, sees computers changing our identity in the direction of more versatility, flexibility, instability, and multiplicity and celebrates this transformation.

According to Turkle, as computers have transformed so did human identity. Today human identity is very different from the one produced under modernist conceptions of computing. Under modernism, the computer was seen as a "calculating machine," as the essence of a "linear, logical model," a model that, according to Turkle, also prevailed in the discourse of "economics, psychology, and social life" (Turkle 1996). This conceptual framework has now changed: "Programming is no longer cut and dried," and it is no longer about "calculation and rules." Instead, "The lessons of computing today have to do . . . with simulation, navigation, and interaction . . . Today's computational models of the mind often embrace a postmodern aesthetic of complexity and decentering" (Turkle 1996).

Through the interaction of humans with computers, these technical and technological changes now bring to bear on the self-conception of humans as cyborgs and on their identities as networked and distributed. Referring to the culture of MUDs (multiuser dungeons—a type of multiplayer online game and one of the sites for Turkle's empirical studies), Turkle says,

As players participate, they become authors not only of text [which is how the game is played] but of themselves, constructing new selves through social interaction. Since one participates in MUDs by sending text to a computer that houses the MUD's program and database, MUD selves *are constituted in interaction with the machine.* Take it away and the MUD selves cease to exist: "Part of me, a very important part of me, only exists inside PernMUD," says one player. Several players joke that they are like "the electrodes in the computer," trying to express the degree to which they feel part of its space. (Turkle 1996, emphasis mine)

MUDs also give "people the chance to express multiple and often unexplored aspects of the self," and so Turkle argues that "MUDs imply difference, multiplicity, heterogeneity, and fragmentation" (Turkle 1996).

Do Humans Still Exist?

The cyborg in the digital discourse represents a conflation of humans and technology and, in that sense, impinges on the modernist notions of the human as an autonomous entity, a subject and an agent. But, as we have seen above, there is another variant to that narrative where the equation of humans and computers brings up the possibility of a completely "humane" technology—that is, a technology that is human in the modernist sense: reflexive, autonomous, and having (at least the sense of) free will. As mentioned previously, this is not merely an exploration of a science-fictional fantasy world but an opportunity in the digital discourse to ponder what it means to be human in a networked, mediated civilization. That computers can be equated with humans, and may someday be just like humans, brings up a question that the following article asks in the most straightforward manner: "What's it mean to be human, anyway?" (Platt 1995a).

At the background of this question are not only metaphorical comparisons between humans and computers but attempts at constructing intelligent machines, or artificial intelligence: computer programs that are as intelligent as humans. Such attempts have brought up a pragmatic question: how does one decree that such a program is indeed as intelligent as a human? The paradigmatic epistemology, and its corollary methodology in the digital discourse, is the Turing test, named after Alan Turing. Turing, a computing pioneer, suggested that a computer program can be said to be intelligent if its human interlocutor was unable to distinguish this program's utterances from those of a human.

The Turing test and its underlying assumptions regarding the essence of what it means to be human are often contested (see Hayles 1999):

Does *appearing* intelligent to an interlocutor equal *being* intelligent? If a human perceives her software interlocutor to be human, does that necessarily imply that the software is intelligent? Thus, for example, the *Wired* article asks can "a computer use trickery to emulate human responses without being intelligent? What did 'intelligence' really mean?" (Platt 1995a). And vice versa, if humans are administered the Turing test, would they easily and unanimously pass for humans? This brings up all sorts of awkward reflections about the distinction between humans and computers, which undermine the essentialist, modernist foundations of what it means to be human.

These reflections clearly preoccupy the author as he gets ready for an annual contest in search of the most intelligent software. The author will take up the role of a collaborator—an interlocutor that judges would have to distinguish from a computer program—that is, a placebo. As the human collaborators prepare for the test, they are instructed by the contest's organizers: "You must work very hard to convince the judges that you're humans . . . You shouldn't have any trouble doing that—because you are human." But the instructor later goes on to shake their confidence in their inherent humanness: "You are in competition," he tells them, "not only with the programs, but with each other . . . One of you will be presented with an award for most human human. And one of you will be ranked the least human human" (Platt 1995a). What this suggests is that being human (either for a computer software or a human) is a measurable skill (of which one can possess more or less) and not a dichotomous quality (which one either inherently possesses or not). Moreover, humanness is rendered a quality, or a characterization that is independent of the subject commonly referred to as "human," quantifiable, and—as we'll see below—can potentially be either present or absent in humans and in technology. Humanness, in other words, is performative.

As the author fears getting last in the contest and appearing the least human human (possibly even less human than a computer program), he ponders about his humanness, or what we might call his H.Q., his "Humanness Quota": "I'm going to do whatever it takes to seem totally, 100 percent human when we start chatting online. But this raises some weird questions. I am human, so why should I need to fake it? Is it possible for me to seem more human than I really am? And if so, what's the best strategy?" (Platt 1995a). Perhaps this very kind of reflexivity with which he engages might undermine his efforts to look human: "if I try to seem more human, I'll end up seeming less human" (Platt 1995a). Along the lines of postmodern understanding of identity, humanness—like any other identity—is construed in the digital discourse to be constructed, precarious, and performative, not given and essential. Different individuals can be more or less human. For example, the author describes another collaborator in the contest, Linda, as "friendly, spontaneous, outgoing—the absolute antithesis of 'computer geek'." These qualities, he adds, make her a sure candidate "to win the 'most human human' award" (Platt 1995a). This statement also elucidates its opposite: being a computer geek makes a person appear more like a computer program and less like a human. Hence, being human is an identity to be constructed and performed and is malleable to quantitative manipulations.

Given the relatively unadvanced stage of development of artificial intelligence, the contest features a "light" version of the Turing test: the discussion between interlocutors (judges, on the one hand, and computer programs or human placebos, on the other hand) is not open ended but limited to a subject chosen in advance. Topics include classical music, the Rolling Stones, and the authors' topic of choice: cryonics (the freezing of corpses). The first question the author is asked by the judge ("What is the difference between cryonics and cryogenics?") puts him in an awkward position, threatening to expose his precarious humanness: "There's no way I can give a human-sounding answer to a question as dry as this. To seem human, I need to show emotion—but if my emotions are excessive compared with the question, the effect will be false" (Platt 1995a).

To be human, in other words, entails performing as one, convincing others of your humanness, and for that reason one is required to uncover and adhere to the laws that underlie a successful performance. In that sense, both humans and programmers of artificial intelligence engage in a similar project: constructing a human object (or better yet, a simulacrum of a human). Under conditions of disembodiment—with no human body to serve as an identification card for admittance—both can fail. Thus, for example, at that particular year's contest, no computer program was able to convince the judges of being human. But neither did all humans. The aforementioned Linda, described as a "friendly, spontaneous, outgoing" human—was rated by the judges to be the least human among the collaborators. Three of the judges, in fact, mistook her for a computer program (Platt 1995a).

NETWORK UNCONSCIOUS

The chapter thus far has highlighted the dominant narratives of the digital discourse regarding the commensurability of humans and technology, such as cooperation, integration, and assimilation. Nowhere in the digital discourse is this fusion between humans and technology more evident than in the realm of the most private and intimate experiences. In these cases, the integration of technology into the human psyche seems the

most welcomed and complete. It is here, according to the digital discourse, that we can most evidently no longer talk of the utilization of
technology by humans but of full absorption; it is here that humans seem
to "vanish into" computers (Brown 2004). I will discuss this narrative of
the digital unconscious through the analysis of two items in *Wired*.

TECHNOLOGICAL MASTERY OF DREAMS

The artwork in "Found" (Kaplan 2005) depicts a person sound asleep.
A banal, quotidian scene complete with soothing bluish bedding, a
bedside table, and an electric alarm clock. But once again, as in "Time
Warp" mentioned previously (Brown 2004), this familiar scene is a setting for a much more disquieting posthuman fantasy. Attached to the
man's forehead are two electrodes presumably connected wirelessly to the
radio on the bedside table. This, however, is not a regular radio, but a
device, manufactured by Sony, called a "Dreammachine." It is in fact
a computer that runs "dream scenario" programs in the sleeper's mind.
In this case, the person has chosen to run a "romantic thriller" (Kaplan
2005). The dreamer not only chooses the general story to be immersed
in but also controls various variables within the dream. In this case, for
example, he chose to have Scarlett Johansson and Elijah Wood as "celeb
avatars"—that is, as protagonists in his dream. He also chose to experience the events of the dream from the "viewpoint" of a "first person," to
set the "action level" to medium and the "titillation level" to high, and
to turn off "parental control." Finally, the device allows the sleeper to set
the level of "emotional intensity" (Kaplan 2005). In addition to running
the dream program over the sleeper's mind, it is clear from the diagnos-
tic meters on the device (featuring an electroencephalogram reading and
determining the sleep stage) that the device is interactive and can read the
vital statistics of the sleeping person.

This is a fantasy, of course, but a fantasy of *what*? What can it tell
us about the ontology of the network human in the digital discourse?
In this fantasy, the postmodernist, poststructuralist undertones are
evident, involving the withering away of two structuralist assumptions: that signs stand for real referents and that there is a distinction
between subjects (or subjective and internal experiences) and objects
(or objective and external realities). In this fantasy scene one can control one's dreams. Rather than expressions of the psyche, an arena for
the unconscious to surface, dreams are here construed as entertainment
brought about by information technology, spectacles to be *screened* and
watched. In the modernist and structuralist framework, dreams are seen
as the epiphenomenal signs of a much more stable and deep human

essence (such as the unconscious). As such, dreams are an *internal* realm where anxieties, unresolved conflicts, and other unconscious forces are reflected and can therefore also be worked out and resolved.

In this fantasy, dreams do not stem from some internal reality. Instead they are seen as free floating symbols, a simulacrum, standing for nothing but themselves. The sleeping protagonist in the photo still experiences a dream in the form of visual signs, but in this case these are external signs imported from without. If computers and the unconscious, network technology and mind, are essentially one and the same thing—computational hardware that produces informational software—what difference does it make where dreams come from? If both are information processing machines, what difference does it make what the source of one's dreams is? In other words, the digital discourse cannot define a qualitative, substantive difference between the dreams produced by human unconsciousness and those managed by the "Dreammachine." In fact, the only difference becomes quantitative: which of these two informational processing machines does it better? As the *Wired* fantasy clearly shows, on this leveled playing field the computer is superior to the human mind in allowing more control over the spectacle and the emotions it evokes and the control over dream scenarios, protagonists, and events. It renders dreams a truly interactive experience.

In this fantasy the external domain of network technology, or cyberspace, penetrates humans' most intimate and personal spaces through information. (In this context it seems significant that the subject matter of the protagonist's dream is clearly sexual. Given that in the modernist Freudian discourse sexual content predominates the unconscious, this makes the statement in the artwork doubly poignant: rather than seeing sex as a foundational motor of human behavior, and hence as a key to self-understanding, sex in the artwork is construed as a spectacle, a tool of satisfaction, and a commodity for consumption). In the digital discourse, cyberspace is seen not as an external entity to the human self but as constitutive of it, specifically, because of its capacity to augment experiences. Network technology helps construct or produce these experiences—and in turn, construct the human self—in ways that offer more control and interactivity and are therefore more empowering. No longer slaves of their Freudian Ids, humans are now promised technological mastery over their unconscious. The penetration of network technology into our innermost spaces is therefore welcome with open arms.

As this occurs, two corollaries follow, uncovering other aspects of the digital discourse on the new network human. As the dichotomy between internal and external is dismantled, dreams become prepackaged, standardized, and commodified just like movies and video games. One can therefore buy a dream, just as the protagonist in the artwork chose to purchase a "romantic thriller." The second corollary is the rationalization of sleep (and dream) time. One of the data discs laying around the side table in the artwork is titled "Learn Farsi in 7 Nights" (Kaplan 2005). If dreams can be reconstrued as the internal processing of (external) information, one might as well process other forms of information. Sleeping hours, the scene suggests, can be used not only for entertainment but also for more industrious tasks, such as learning a foreign language. To recall a theme developed in Chapter 4, the blurring of the distinction between work time and leisure time is here even more complete and intimate; through the penetrating capabilities of network technology, even sleep time can be instrumentalized and rationalized; even the unconscious can be rendered a space of production.

INTERNAL EXPERIENCE THROUGH EXTERNAL MEANS

The triumphant narratives about the possibilities of augmentation and rationalization of human experience through information technology are succinctly captured in another piece in *Wired* magazine, this time expressed as a prediction rather than a mere artistic fantasy from the eminent filmmaker Steven Spielberg. Prefacing an interview with Spielberg is a two-page illustration and an epigram featuring a quote from the interview. The epigram reads, "Someday the entire motion picture will take place inside the mind. It will be the most internal experience anyone can have" (Kennedy 2002). What is striking about this quote is not simply the prediction that someday such a technological feat will be possible, and not even that such development is seen as positive and desirable—as such, it is yet another expression of the fantasy of total immersion of technology and human mind. What is particularly striking here is Spielberg's use of language, deeming such an experience, should it occur, an "*internal experience*" (Kennedy 2002, emphasis mine).

Spielberg's use of the concept of "internal" to refer to such an experience is quite counterintuitive, since the source of the experience—by Spielberg's own logic—is clearly *external* to the mind, as the interview and the illustration that accompanies the epigram make clear. In the interview, Spielberg is asked about the future of technology and storytelling. He says in reply, "The technology may give us far better tools to communicate our stories. The technology may also provide a theater of the mind. Someday the entire motion picture may take place inside the mind, and it will be the most internal experience anyone can have: being told a story with your eyes closed, but you see and smell and feel and interact with the story" (Kennedy 2002). Spielberg leaves no room for doubt: the source of what he predicts to be the *utmost* internal experience—like a dream, perhaps—is external; the mind will feel what the computer projects.

The illustration that accompanies the epigram is even more revealing, further uncovering the blurring between "internal" and "external" in the digital discourse. It features a schematic profile of a human head, the likes of which is frequently used to educate about the neurophysiology of sight. Taking the eyeball as a starting point, we see a reverse still image of an iconic scene from the movie *Casablanca*. At the other end, on the brain, or the mind, is a "projection" of the same image upside down and much enlarged. This would be the conventional direction of the mind processing an external image. But what this illustration suggests is that "seeing," according to the futuristic scenario painted by Spielberg, is merely a reflection of the images coming not from an external source but from the brain (the reverse direction is indicated by the spiraling arrows). By any measure, "seeing" a movie this way would also be a very *external experience*, since the signs will still be produced externally—by a storyteller, a programmer, and by information technology—and transmitted to the human mind.

Spielberg's word choice is not a mistake; quite the contrary, it has been chosen by *Wired*'s editors to adorn the two-page introduction to the interview precisely because it is such compelling language. That Spielberg is able to denote such experience as *internal* is a testimony to a breakdown of the very distinction between internal and external in human experiences in the digital discourse. Such modernist language seems incompatible with the new subjectivity and its new experiences, inextricably tied with information technology. As such, these experiences are at one and the same time external and internal, taking place at the technological and the psychological realms. For what does it mean for the network human—a conflation of human and technology—to have an internal experience if his experiences are processed, transmitted, and perceived by both mind and machine, if his identity and consciousness is distributed among several computational machines? His existential sensibilities, lived experience, and identity become intricately bound with technology; his innermost experiences are rationalized and technologized by external machines.

CONCLUSION: BETWEEN SELF AND THE NET

According to the digital discourse, in the information society, humans can no longer be understood as separate from technology. From the molecular level to the realm of lived experience, from the body to psychic life, from the human mind to cyberspace, human beings in the most technologically advanced regions of the world have been inextricably fused with network technology. Under these new circumstances, humans can no longer be thought of within the framework of humanism as distinct from technology but instead as network humans. Network humans are commensurable with network technology. This means that they are conceived in terms of network technology, that they are able to get into intimate relations with them, and, finally, that network technology resonates with the true essence of humanity.

The commensurability between humans and network technology is made possible, according to the digital discourse, because of the revolutionary nature of network technology. The underpinnings, characteristics, and architecture of network technology—such as binarism, information, communication, flexibility, and adaptability—echo the essence of humans. This commensurability allows for new forms of interaction and convergence between humans and network technology to take place: humans can be further technologized and instrumentalized, technology can be humanized, and the hybrid entity of the cyborg can be constructed.

Since the mutual transference between humans and network technology is rendered unproblematic in the digital discourse, network technology is also seen as more conducive to bringing out human qualities associated with the humanist critique of capitalism: underdetermination (or nonessentialization), flexibility, and multiplicity. In the digital discourse, the conflation of information technology and humans, conceptually and practically, is seen as a means for human augmentation and emancipation. It is not simply that to be human is inextricably tied to information technology; human experience is augmented by technology, and human potentials are unleashed. Since network technology and humans are perceived as having a shared essence, network technology is seen in turn as the ideal conductor for human's liberating potentiality.

These narratives are weaving a spirit of networks for a post-Fordist society. Humans are reconceptualized to fit their place in the new mode of production and a new society: contrary to the Fordist human characterized by spatial and temporal bodily presence and by physicality, the post-Fordist human is characterized by virtuality and disembodiment. The post-Fordist human—her body, mind, and identity—are informational, flexible, and multiple.

Set within the framework of this book, the digital discourse on the network human is seen as a response to the humanistic critique of Fordist society concerning authenticity. It rejects and transcends the overdeterminist and essentialist conception of the human in Fordist society. Network technology is conceived as materializing a response to the humanist critique of capitalism regarding authenticity, creativity, and self-expression; network technology complements the network human. In the digital discourse the relationship between humans and network technology (resulting in the network human) is set in contrast to the relationship of humans and industrial technology (resulting in industrial human).

FROM INDUSTRIAL HUMAN TO NETWORK HUMAN

The construction of the industrial human during Fordism—while allowing smoother integration of humans into the industrial machine—is seen in retrospect in the digital discourse as a hindrance to human emancipation and growth because it did not resonate with the true essence of humans but, in fact, suppressed it. During industrialism, the flexibility of the human body, mind, identity, and unconscious encountered stifling machines. The interaction of humans with machine followed the rigid rationale of command. It was not, in fact, interaction at all. In the best case, humans utilized technology by command, and, in the worst case, humans themselves were co-opted into the technology, becoming cogs in the system. With network technology humans have a much deeper interaction, which results in fusion, integration, and merging, and in the construction of a new subjectivity. The blurring of the boundaries between humans and network technology allows for a more meaningful, emancipatory, and natural interaction of humans with technology and unleashes the emancipatory potentials of humans.

We should recall here how the disciplining of humans to industrialism in general and Fordism in particular has been conceived by Gramsci. It is precisely this sort of analysis at the beginning of Fordism that the digital discourse continues and responds to (short of Gramsci's critical thrust). It is precisely Gramsci's concerns that are presumed in the digital discourse to be overcome by network technology. It is therefore quite instructive to juxtapose the digital discourse's construction (and celebration) of the network human with Gramsci's critique of the industrial human, a reading that reveals the extent to which the network human is constructed as a response to the industrial human. In "Americanism and Fordism" (Gramsci 1971), written originally in the 1930s, Gramsci says that,

> the history of industrialism has always been a continuing struggle (which today takes an even more marked and vigorous form) against *the element of "animality" in man*. It has been an uninterrupted, often painful and bloody process of *subjugating natural* (i.e. animal and primitive) instincts to new, more complex and rigid norms and habits of order, exactitude and precision which can make possible the increasingly complex forms of collective life which are the necessary consequence of industrial development. *This*

struggle is imposed from the outside, and the results to date, though they have great immediate practical value, are to a large extent purely mechanical: the new habits have not yet become *"second nature."* (Gramsci 1971, 298, emphasis mine)

At the heart of the creation of a new industrial human, Gramsci identifies a process of suppression of humans' natural disposition by external means. The digital discourse on the network human directly responds to these concerns by suggesting that network technology responds to, and augments, the already-existing "first nature" of humans as flexible, informational, multiple, and so forth. What is more, with network technology this "struggle," as Gramsci puts it, is no longer imposed from the outside, since the ecology of network technology resonates with humans' internal characteristics. The network human offers a new subjectivity that transcends the boundaries between humans and technology and, in turn, transcends the limitations of the Fordist human. In the process, technology and technological reason are even more deeply internalized into our understanding of what humans are, and into human practice.

Unlike the digital discourse's celebration of the network human, Gramsci points to the social coordinates underlying the process of technologizing humans both discursively and practically. Referring to the forefather of scientific management, Gramsci says, "[Frederick] Taylor is in fact expressing with brutal cynicism the purpose of American society—developing in the worker to the highest degree automatic and mechanical attitudes, breaking up the old psycho-physical nexus of qualified professional work, which demands a certain active participation of intelligence, fantasy and initiative on the part of the worker, and reducing productive operations exclusively to the mechanical, physical aspect" (302). Here again, the digital discourse's narrative regarding network technology as unleashing "active participation of intelligence, fantasy and initiative" (302) in the network human is further elucidated.

Gramsci's analysis of Fordism elucidates a critical point: humans are constructed and transformed through the productive process and through the technology that dominates this process. (It is no surprise, then, that through the discourse on technology human nature is also "uncovered.") It is transformed in such a way that renders humans malleable to and compatible with the new productive process and the broader social arrangements it entails.

Fordism entailed a new approach to (and a new technology of) the labor process—indeed to the technologizing of the labor process—that Henry Ford fathered and implemented. Ford's introduction of the assembly line, together with the introduction of Frederick Taylor's scientific management, revolutionized the labor process by creating a division of labor within the factory. Taylor (1967) intensified the process of the rationalization of production by transforming the continuous process of work and the embodied character of skills into discontinuous and disembodied segments. Those autonomous units could then be individually perfected (first by creating an increased division of labor and, ultimately, by automation), as well as rearranged over time and space. Taylor's idea was to abstract labor by (a) treating humans as machines (in the positive sense of the term, that is, as programmable, measurable, adaptive, malleable, and so forth) and improving, or technologizing their physical skills as workers; and (b) reformulating the whole process of work as information so that it can be completely rationalized and be performed by machines (Robins and Webster 1999, 87–108; Mattelart 2003, 37–40). Mass production and mass consumption, the large-scale bureaucratic corporation, and the deepening rationalization of the work process—staples of Fordism—have been enabled by technology and—to no lesser degree—with the reconceptualization of humans in accordance.

Thus, for example, in contrast to the digital discourse's discovery that humans are essentially about information, during Fordism, it was *energy* that was constructed (or "discovered") as the commensurability unit between humans and technology. With Fordism, we witness the emergence of what Anson Rabinbach calls *The Human Motor* (1992). Rabinbach shows how the emergence of the discourse on the human motor was conjoined with the dominant scientific and technological transformations of the time. Helmholtz's discovery of the laws of energy as the underlying operational rules of the universe and their construction as universal and abstract entities were central to a revolution in the perception of humans as machines for the extraction of labor power. This discourse had led to the emergence of a new science of labor—the most renowned practitioner of which was Taylor—that in turn led to new bodily practices involved in the labor process. Rabinbach is pointing out to a dialectical relationship, which results in a new social totality:

> With the invention of the steam and internal combustion engines . . . the analogy of the human or animal machine began to take on a modern countenance . . . [T]he eighteenth-century machine was a product of the Newtonian universe with its multiplicity of forces, disparate sources of motion, and reversible mechanism. By contrast, the nineteenth-century machine, modeled on the thermodynamic engine, was a "motor," the servant of a powerful nature conceived as a reservoir of motivating power. The machine was capable of work only when powered by some external source, whereas the motor was regulated by internal, dynamic principles,

converting fuel into heat, and the heat into mechanical work. *The body, the steam engine, and the cosmos were thus connected by a single and unbroken chain of energy.* (Rabinbach 1992, 52, emphasis mine)

The emerging discourse on the industrial human at the beginning of the twentieth century was intertwined with practices that in effect created a new type of human, of which Gramsci talked. Rabinbach tells of how "the emergence of a physiological approach to labor coincided with important changes in work during Europe's second industrial revolution," which involved the introduction of "electric power, of steel and chemical steel and chemical production, and of the rise of industries producing heavy machinery" (Rabinbach 1992, 122). The new factory introduced new types of workers (unskilled immigrants, leading to rapid turnover) and a new labor process that eliminated craftsmanship and was determined by new technologies (Rabinbach 1992, 123; see also Noble 1984). Discourses, as Foucault has taught us, are practices "that systematically constitute the subjects and objects of which they speak" (Schwandt 2001, 58).

THE INTEGRATION OF HUMANS INTO NETWORK CAPITALISM

This theoretical framework and historical precedents should encourage us to highlight the compatibility of the network human to contemporary capitalism. In light of the detailed description of the digital discourse in the previous chapters of Part II, these should render the compatibility almost self-evident: a network market, network work, and network production require the reconceptualization of humans as commensurable with network technology as informational, flexible, and distributed and as nodes in a techno-human network.

One of the corollaries of the network human's inessentiality and flexibility is the understanding of humans as constructed, indeterminate, unstable, and multiple. These can be conceived as prescriptive characteristics for successfully living in the network (and for becoming, in Boltanski and Chiapello's terms, "a great man," [2005, 112–21]), as they are a description of the true essence of humans. This can be exemplified with the changing work ethics of the new capitalism. The new, flexible, and networked capitalism demands workers to constantly move between multiple tasks and ad hoc projects that demand multiple and varying skills. The old work ethics of apprenticeship and ever-deepening craftsmanship in a single skill is therefore replaced by an ethics of superficiality and multiplicity (Sennet 2000, 98–117, 2006, 83–115). Under these new circumstances, characteristics such as "proficiency" and "understanding" receive a new meaning. In the aforementioned article by Sherrie Turkle, she introduces us to the subjective experience of thirteen-year-old Tim, a gamer of the virtual reality game SimCity: "The goal [of the game] is to make a successful whole from complex, interrelated parts . . . The game is about making choices and getting feedback . . . [about] zoning restriction and economic development, pollution controls and housing starts" (Turkle 1996).

Tim, in other words, learns multitasking, which involves working in multiple sites with only superficial understanding and minimal skills. Tim, says Turkle "is able to act on a vague intuitive sense of what will work even when he doesn't have a verifiable model of the rules underneath the game's behavior . . . Tim can keep playing even when he has no idea what is driving events." Note the following exchange between Turkle and Tim, after she asks him why one of the species of his virtual planet became extinct:

Tim: I don't know, it's just something that happens.
Turkle: Do you know how to find out why it happened?
Tim: No.
Turkle: Do you mind that you can't tell why?
Tim: No. I don't let things like that bother me. It's not what's important."
(Turkle 1996)

After the game sends Tim a message, "your orgot is being eaten up," Turkle asks him "'What's an orgot?' He doesn't know. 'I just ignore that', he says. 'You don't need to know that kind of stuff to play'." Later Tim has to appease Turkle again, saying "Don't let it bother you if you don't understand. I just say to myself that I probably won't be able to understand the whole game any time soon. So I just play it" (Turkle 1996). Turkle celebrates Tim's approach as a sign for a new liberated human; after all, humans are natural multitaskers, adaptable, flexible, and pragmatic.

The compatibility of the network human to network capitalism is evident also in the discourse on prosumption. Chapter 5 registered the construction of prosumption (as a process) and prosumers (as a social category) and elucidated the significance of these in the context of the new capitalism. The discourse on the network human further cements these categories with the narratives pertaining to the commensurability of humans and network technology. Commensurability interjects humans into the loop of production and consumption in a radically intimate fashion. The network human is in fact constructed as a site of prosumption. Through commensurability and distribution, the body, self, and identity are reconstituted as sites not simply of independent practices of production and consumption but of prosumption (Lowe 1995). Such an intimate human-technology loop is described, for example, in the afore-mentioned *Wired* article on Mattew Nagle (Martin 2005). The implica-tion of the blurred boundaries between humans and network technology is that intelligence, creativity, self, identity, and experience are not limited to the demarcated structure of the human but overflow to the network.

The list of compatibilities of the network human to the network capi-talism could go on: the disembodiment and distribution of the network human (i.e., the separation of thoughts, feelings, and experiences from the body) and network capitalism's demand for work to be performed every-where; the network human's fragmentation and flexibility and the recon-ceptualization of worker not as a whole but as a part of a business project; and the underdetermination and instability of network identity and the demands for constant flexibility and adaptability to system's needs.

Reconstructing the narratives of the information revolution through an examination of the "visions," or *Technomanifestos* (Brate 2002) of some of its protagonists, such as Norbert Wiener and Alan Turing, Adam Brate (2002) notes that "the goal of the information revolutionaries [was] to create new systems—technological, social, political, and economic—that adapt to people instead of the other way around" (Brate 2002, 4). Put in the language of this book, the narratives that Brate registers assign network technology with the purpose of creating a new human that tran-scends the industrial human. In a similar fashion, albeit critically, Jodi Dean sums up this project thus: "Technoculture is an ideological forma-tion that uses democracy, creativity, access, and interconnection to pro-duce the subjectivities of communicative capitalism" (Dean 2002, 103).

Informationalization, disembodiment, distribution, fragmentation, inessentiality, instability, flexibility, interactivity, play—according to the digital discourse, it is precisely these authentic human potentialities that industrial-mechanical technology suppressed and can now be unleashed and flourish with network technology. These are also the human charac-teristics that, as Dean (2002) points out, render the new kind of network capitalism possible.

PART III

CHAPTER 7

NETWORK COSMOLOGY AND THE EXHAUSTION OF CRITIQUE

PART II OF THE BOOK OFFERED AN ANALYSIS OF the digital discourse pertaining to the intersection of network technology with key sites of the new capitalism: the market, work, production, and the human. It underscored how the digital discourse constructs a spirit of networks that legitimizes new constellations of power, new modes of work, new relations of production, and new ways of being. Part III of the book offers two concluding discussions that highlight two theoretical and sociological dimensions of the digital discourse. Chapter 7 registers and explains the *technologistic* facet of the discourse on network technology. It further asks, "What is the social meaning of a discourse that revolves on technology as its axis of understanding and presentation?" Chapter 8 summarizes the main findings of the book, reiterates its central arguments, and delves into the ideological and political dimensions of the digital discourse as the new spirit of network capitalism.

After the four previous chapters examined central loci of the digital discourse concerning the new capitalism, we turn now to another level of analysis. Here, I focus not on the *socially informed* themes of the digital discourse but on the *technologically informed* framework with which these themes are analyzed. To use a metaphor, this chapter focuses not on the bodies of knowledge on which the digital discourse operates but rather on the scalpels it is using. The chapter cuts across the substantive themes of the previous four chapters in order to uncover the underlying epistemological coordinates of the technologically centered discourse of the digital discourse. As such, it also serves as one of the two concluding chapters of this book.

This chapter examines the digital discourse on what I call the cosmology of network technology, that is, the discourse on the ontological, historical, and social status of network technology itself. We have seen network technology starring in the leading role of almost every article in *Wired*. It is, as I have indicated in Chapter 2, the *raison d'être* of the magazine: to capture the world through the prism of network technology. Underlying this description of the world is a particular theory, or a set of assumptions about the nature of network technology, the relations between technology and society, and most importantly, about the nature of society itself at the beginning of the twenty-first century. The purpose of this chapter is to present and analyze this theory and uncover the assumptions about network technology that underlie and inform the digital discourse's analysis of the digital society.

At the foundations of such cosmology is an answer to "the question concerning technology," as Heidegger puts it (1977): what is technology, or, more sociologically speaking, what are the relations between technology and society? The answer—both in the social sciences and in public discourse—has overwhelmingly been that technology is the chief agent of social change, that is, that technology makes society (Murphie and Potts 2003, 11). This view—termed by Robins and Webster as "technologism" (1999), that is, making technology the axis of social explanation—makes three assumptions:

1. *Neutrality*. Technology has a history of its own, and its development stems from an internal dynamic; in that sense, it is an asocial force, external to social power struggles. The development of technology is seen as stemming from an internal logic. Technology comes into being in a sphere external to society. This is a *Machina ex Deus* approach, which perceives technology to be the love child of technology, stemming from thin air or the otherwise nonsocial. The genius inventor, struggling in his (and, rarely, her) laboratory to come up with an invention—usually for the sake of innovation itself—is an epitome of that perception. According to that, in order to understand technology we need to adopt an "internalist" approach (Bijker 1995, 9–10), that is, we need to examine its minute technical details. Technology is fetishized (Robins and Webster 1999, 51–53) as a force completely autonomous from society (Webster 1991, 1). Since technologies are conceived as having their own autonomous functional logic they can therefore be explained and analyzed without reference to the "external," or social factors (Feenberg 1995).

2. *Inevitability*. Technology largely and unidirectionally determines the shape of society, reconfiguring it in accordance with its internal working; once a technology is presented and adopted by society, it generates an outcome, which is inscribed in the essence of that technology. It might not always be forecasted, but social change can always be

traced back to technological factors. According to this proposition, technology is an autonomous entity (Winner 1977) that drives history (see Smith and Marx 1994). Thus, technology is perceived as a *fait accomplis*, something to which society has to adopt passively. The reaction to technology can only be conceived within a dichotomous framework of "love it or leave it" (Feenberg 1991).

3. *Benevolence.* Technological progress is human progress; it is positive and benign in and of itself. Some formulations go so far as suggesting that only technology can promise human and social transcendence; progress *is* technological advancement, hence the longstanding strand of technological utopianism in the most technologically advanced societies (Segal 1985). This makes critical views of technology especially difficult to voice, since they are automatically framed as conservative, romantic, and reactionary, or—the designated derogatory—Luddite (Robins and Webster 1999, 39–41; Postman 1993, 43).

This chapter presents the digital discourse on network cosmology along three narratives. These narratives both echo the tenets of technologism in general and update them to the specificities of network technology. The digital discourse, the chapter shows, naturalizes, theologizes, and teleologizes network technology. It argues that network technology is more akin to nature than any previous technological paradigms. With the added premise that technology largely determines society (so that technological characteristics are inscribed in society), it creates a trinity whereby society itself is presumed to take on a more natural form with network technology being the medium between the natural and the social.

Related to the rhetoric of naturalization is the idea that technology offers a form of transcendence almost spiritual and religious in nature. According to this theological narrative, the increased digitalization and networking of social life (i.e., its rationalization) does not entail a process of *disenchantment* (the hallmark of rationalization during industrial technology) but rather offers a road to transcendence. As Weber's "iron cage" (Weber 1958) becomes digitized and networked, it also wins back the soul it had lost with the emergence of industrialism. Lastly, the digital discourse constructs a cosmology in which humans recede to the background and occupy a supporting role in a new history, where technology is the protagonist. In such a history, network technology is seen as the teleological climax not simply of the history of technology but of the history of the universe. It is in this sense that technology is teleologized, that is, becomes the subject of its own history, and is dissociated from the social context within which it is developed and deployed.

NATURALIZING TECHNOLOGY: THE TRINITY
OF NATURE, TECHNOLOGY, AND SOCIETY

The idea that technology is the prime motor of social change is a cornerstone of technologism. It has certainly been a central thread in technology discourse throughout modernity; indeed it has been a central thread of the discourse *on* modernity. But in the digital discourse, this argument is made even more forcefully, upgraded from a general theory of technology, and suggested as the quintessential characteristic of network technology *in particular*.

According to the digital discourse, network technology is truly socially transformative, since it resembles the workings of nature more than any technological paradigm before it, specifically, more than industrial technology. By replicating the ways nature works, network technology serves as a bridge between the natural and the social, hence transforming social dynamics in ways that accord more with nature. This technology-nature theory suggests that the binary language on which information technology is founded complies with the laws of nature. In other words, the process of informationalization—that is, the process whereby the material and social world are translated into digital language and hence integrated into information technology—is not a capricious imposition of some technological or mathematical invention on the world but a long-overdue uncovering and return to the natural order of things.

In the same vein, the architecture and operational modes of the network and the communicative and organizational forms that sprout from it—such as chaos, spontaneous order, decentralization, and flexibility—are all seen as natural insofar as they replicate the workings of nature. To the degree that nature has been reified and depoliticized through much of modernity—that is, constructed as lying outside of human affairs—the association and conflation of network technology with nature is meant to depoliticize technology by quite literally naturalizing it. Table 7.1 summarizes the abstract logical assumption at the heart of the naturalization argument of the digital discourse.

It is therefore commonplace in *Wired* to conflate technology and nature, to find confirmations for one by its homologous affinity to the other, and to suggest that information technology is natural and, vice versa, that nature is informational. Hans Moravec, a researcher at Carnegie Mellon's Mobile Robot Laboratory and the author of *Robot* (1999), goes a long way in converging biology and technology, predicting that "mechanical minds are about to replay the evolution of biological minds" (Davis 1998). The development of information technology, Moravec suggests, got on track with biological entities and their evolution, and natural

Table 7.1.   The logical assumptions of the technology-nature-society thesis

Network technology follows the laws of nature	$NT = N$
Network technology shapes society	$NT \dashrightarrow S$
Therefore, society works more naturally	$S \approx N$

evolution is now replayed in the technological domain. In the same vein, in a review of a book on the history and future of the information revolution—*The Soft Edge: A Natural History and Future of the Information Revolution* (Levinson 1998)—William Goggins also likens the evolution of media to biological evolution. The book, he writes in a McLuhanite language, suggests how newer communication technologies "outpaced their ancestors' limitations, gradually extending human faculties across space and time" (Goggins 1998).

Life Is Information

But the conflation of network technology and nature is more than casual and metaphoric, as evident from "Revolutionary Evolutionist" (Schrage 1995), an exposition to the work of Richard Dawkins, the author of the best-selling *The Selfish Gene* (1976), an oft-cited book on the pages of *Wired*.[1] I will leave the "selfish" aspect of the argument aside and focus here on the nature and technology nexus. According to the article, Dawkins's book argues that "living things are little more than corporal vessels impelled to heed the primal dictates of selfish genes hellbent on their own replication and propagation." Taking this reductionism a little further, Dawkins argues that "genes themselves are expressions of particularly elegant *code* manipulating the world around it to its own productive end" (Schrage 1995; emphasis mine).

Life's dynamics, Dawkins argues, are determined not at the level of "animals," "beings," or "society" but at the much simpler and smaller level of genes, which are nothing more than an informational code. "All life," he argues, "at its core, is a process of digital-information transfer" (Schrage 1995). Information and communication are the backbone not simply of computer programs, social life, or global financial institutions; they are in fact the backbone of life itself. Dawkins's theory shifts the focus "away from the individual animal as the unit of evolution" to the gene and to the group (or network) communication of genes. He concludes that the way genes behave over time "is the very definition of an evolution based on the flow of information" (Schrage 1995). The article hails Dawkins for redefining "the fundamental doctrines of 'natural selection' in ways that *transform the vocabulary of evolutionary biology into the new realms of digital media*" (Schrage 1995; emphasis mine).

Dawkins considers his theory, which in some respects takes after Darwin's, to be the ultimate explanation of life; that is, of both biological and social life, which are really surface phenomena of a deeper, informational code. Most significant for our discussion here, "Dawkins has been extremely effective in probing the boundaries between natural evolution and artificial evolution as created in computers" (Schrage 1995). The very distinction between natural and technological evolution is, according to Dawkins, artificial and redundant. Artificial life researchers, for example, believe that "life is an information process that can be ported from one matrix to another." Advancements in information technology seem to facilitate such exploration and reaffirm the nature and technology nexus: "The rise of cheap processors and parallel architectures creates the ideal digital ecosystems to spawn software rather than build it. Nature—not rational cognitive planning—becomes the guiding force for the next generation of software solutions" (Schrage 1995).

Since genes cannot sufficiently explain the complexity of human life, Dawkins's theory introduces another concept, "memes," which are ideas, or other informational codes, that "are to cultural inheritance what genes are to biological heredity." Human evolution can be explained as the coevolution of genes and memes. "This idea," the author suggests, "offers a powerful intellectual framework for a new understanding of life as an information process" (Schrage 1995). In this framework, humans are redefined as mediums of coded information and its communication. The real protagonist of human social history is that which is at the heart of natural history, which in turn is exactly that which is at the heart of the digital revolution: informational code.

This, we should note, is not simply a matter of metaphors or semantics. Since models and metaphors are really all we have to grasp the world, they very much determine our view of the world, the associations we make, and the impressions that we are left with. The associations of technology with nature, of technological advances with natural history, and of digital media with natural selection reaffirm this technologistic point of view.

TECHNOLOGY AS A NATURAL FORCE

Like Dawkins, Kevin Kelly, too, equates network technology with nature and, in fact, converges natural and technological history. In Chapter 3 I dealt extensively with Kelly's discussion regarding the social forms of the network society. Here I will focus on the technological assumptions on which these social forms are presumed to be predicated. Examples are taken from Kelly's *New Rules for the New Economy* (1998). Right at the outset Kelly asserts the centrality of technology in society: "No one," he says in the opening sentence of the book, "can escape the transforming fire of machines" (Kelly 1998, 1), and he therefore suggests applying a simple rule to understand social and economic reality: "Listen to the technology . . . find out what it is telling you," he says (8), upholding what Bijker identifies as the "internalist" view on technology (Bijker 1995, 9–10). The socially transformative trajectory of technology, Kelly implies, lies in its internal essence and history.

Network technology seems to be telling Kelly about nature. To grasp the enormity of the contemporary technological revolution, Kelly situates it on a historical axis that begins a few billion years ago: "It took several billion years on Earth for unicellular life to evolve. And it took another billion years or so for that single-celled life to evolve multicellular arrangements" (Kelly 1998, 6). But these multicellular arrangements, he explains, yielded very limiting forms of life, since cells had to be directly connected to each other in order to work in coordination: "After another billion years, life eventually evolved the first cellular neuron—a thin strand of tissue—which enabled two cells to communicate over distance" (6). This, he says, was a quantum leap in the possibility to create more diverse and sophisticated forms of life. In a language that echoes the biblical scene of creation, Kelly recounts, "Life quickly exploded in million different unexpected ways, into fantastic awesome varieties, until wonderful life was everywhere." After this short detour into natural history Kelly makes a clean break back to technology and concludes, "Silicon chips," he says, "linked into high-bandwidth channels are the neurons of our culture" (6).

This is not simply a metaphor, a projection of a natural science discourse over a technological screen. For Kelly, the silicon chip is an extension of living organisms and squarely fits into the history of the natural world. Previous chapters of this book paid close attention to the actual dynamics and forms attributed to the network in the digital discourse—complexity, decentralization, spontaneous order, and so forth. What we need to pay attention to here is the epistemological level of the argument, in which (a) information and communication are defined as the essence and teleology of nature and (b) a direct, unproblematized, and undialectical transference between nature, technology, and society is made.

The transformative power of technology, Kelly says, is true for any technological paradigm, like the steam engine or electricity, and the same goes for network technology. The reason why Kelly is ready to put so much weight on network technology as transformative and why he is ready to surrender completely to technological determinism—to embrace it, in fact—is because he is able to conceive network technology as the end of technological history. Not, of course, that this ends the cycle of innovation; quite the contrary, technology, its improvement, and further application to every sphere of life are central to the digital discourse. Rather, he sees in network technology the final *paradigm* of technology. That is because technology—a human creation, a product of human culture—has finally took the shape of nature itself. The paradigm of network technology inaugurates a science and technology that follow the rules of nature, replicating the way biological systems work, sustain themselves, and evolve. After taking a long detour from these natural rules (emblematic to the unnatural technological paradigm of the industrial era), human civilization has come full circle, finally joining the bandwagon of nature.

Network technology is benign, transformative, progressive, and inevitable, according to the digital discourse, not simply because it affects human life in positive ways but, more fundamentally, because it is natural, that is, it follows the way things work in nature. Its transformative capacity ensures that society and social formations become better, its structures more adaptable, its potential more fulfilled, and its economy more workable. It also suggests that any intervention in the workings of this technological paradigm is not only futile (the force of natural elements—like a flood, or tornado—is too powerful for humans to resist) but also reactionary. Instead, it suggests that at a social level we should be doing all we can to facilitate these transformations and to accept pitfalls and difficulties arising from network technology as temporary, necessary, and unavoidable obstacles on the way to a brighter horizon.

Unsurprisingly the story of network technology is traced back to nature in the digital discourse. Likewise, the story of nature is told as the story of information and communication; both histories are interwoven. To put it more bluntly, what is traditionally thought of as natural history, dissociated from human history and culture, is retold in the digital discourse as the history of informational codes and network communication, which itself is dissociated from human history. Hence, information and communication become the real teleology of nature, society, and technology. Thus, insights from nature and technology are transplanted to the social field. Our economy, Kelly argues, is only now moving out of the multicellular level: "By the enabling invention of silicon and glass neurons, a million new forms are possible. Boom! An infinite variety of new shapes and sizes of social organizations are suddenly possible" (Kelly 1998, 6). By assuming a direct, causal, and deterministic link between technology and society, and by insisting on their semblance to nature, Kelly constructs a framework where the political dimensions of social structures—in this case, the economy—are uprooted and altogether excluded and where the social is reduced to the technological.

Another theme in the digital discourse that shifts freely between the three levels—nature, technology, and society—pertains to complexity. I have taken up the issue in depth in Chapter 3. What interests me here is again not the substantive argument but its technologistic epistemological underpinnings. Recall Kelly's observation regarding the emergence of dumb chips and their interconnection into a rational smart network. Kelly affirms this by a reference to nature. Since this network model can be found in both nature and in technology, the analogies are interchangeable. In the space of a few sentences Kelly shifts back and forth between the two:

> A personal computer is like a single brain neuron . . . When linked by the telecosm into a neural network, these dumb PC nodes create that fabulous intelligence called the World Wide Web . . . dumb cells in our body work together in a swarm to produce an incredibly smart immune system, a system so sophisticated we still do not fully comprehend it . . . Dumb parts, properly connected into a swarm, yield smart results (Kelly 1998, 13) . . . The surest way to smartness is through massive dumbness (Kelly 1998, 14) . . . Everyday we see evidence of biological growth in technological systems. This is one of the marks of the network economy: that biology has taken root in technology. And this is one of the reasons why networks change everything. (Kelly 1998, 32)

It is obvious that the use of the language of nature and biology to explain technology and society is not simply metaphorical but points to a conflation of these worlds *in reality*. In this case, the social (economy) and the technological (which underlies the networking of the economy) are understood to be overtaken by nature. The crux of this depoliticizing narrative comes from the anchoring of the social in the technological, which in turn is anchored in the natural.

The formidable, socially transformative power of network technology—more than previous technological paradigms—rests, according to the digital discourse, not only on technology being more natural but also on culture being more technological. Kelly makes a distinction between previous technologies and current network technology in terms of their dominance in society. Up until the midtwentieth century, says Kelly, technology was almost entirely separated from society, or, rather, society was autonomous from technology; technologies merely "enhanced the prevailing culture" (32). Technologies could therefore be ignored, since they occupied the periphery of society; they "did not penetrate the areas of our lives we have always really cared about: our network of friendship, writing, painting, cultural arts, relationships, self-identity, civil organizations, the nature of work, the acquisition of wealth, and power" (32–33). But with the advent of the digital revolution, "technology completely overwhelmed these social areas" (33). As Kelly succinctly puts it, "Technology has become our culture, our culture technology" (33).

But how did that happen? Why had technology become so central and embedded in our life? How did it come to be "the campfire around which we gather" (33)? Kelly's answer is that technology "has become more like us. It's become organic in structure" (33). This argument resonates with the themes raised in Chapter 6: by being more organic and natural, technology is able to resonate with humans more fully. The arguments then ultimately circle back to nature: our culture is technological, our technology biological, our biology informational. Technology can no longer be thought of in social terms as something that humans construct and that therefore follows a social logic. Network technology "behaves more like an organism than like a machine" and is much better understood as part of nature (33).

TECHNOLOGY AND THE INEVITABILITY OF NATURE

As David Nye shows in his study of eighteenth- and nineteenth-century technological narratives in America, the naturalization of technology meant that technological—and the ensuing social—transformations were seen as inevitable: "The narratives naturalized the technological transformation of the United States so that it seemed an inevitable and harmonious process leading to a second creation that was implicit in the structure of the world" (2003, 6). In a similar fashion, the naturalization of network technology and the informationalization of nature are reiterated in the digital discourse.

In "The Big Picture" (Hillis 1998), natural history becomes a teleological story that culminates with the advent of information technology. In this long and evolutionary process, "cells evolved an information processing mechanism . . . the genetic code of DNA" (Hillis 1998). Since the biological is understood to be, in essence, informational, information (or knowledge) is portrayed as the protagonist of this natural history: "With DNA came an evolutionary advantage: knowledge, as genetic recipes, could accumulate from generation to generation" (Hillis 1998). Then there was the great transformation from single cells to cells cooperating and creating a community, not unlike a communication and information network. The article goes on to unfold this history until it reaches the time of human civilization with its most sophisticated means of communication—language. Here the ontologies of informational nature and information technology converge and reach their zenith. All sorts of information and communication technologies, the article argues, "are all specialized mechanisms we've built to bind us together" (Hillis 1998). The evolution of living organisms, as well as that of technological machines, has an ontology and teleology—a communication, cooperation, and cohesion of disparate parts. In this framework, network technology is seen as the pinnacle of that evolutionary development and as essentially bent on fortifying natural tendencies. It is thus also essentially benevolent.

In another piece that puts together natural, technological, and social history (Gilder 1998), the author upholds information as the underlying building block of the universe, the key to life: "Ruled by DNA codes, cells are chiefly a symbol system governed by information theory. Codes precede chemistry and are not reducible to it" (Gilder 1998). The foundation of life, according to this, is information, not substance; the secret of life is encapsulates in information, in immaterial codes. This is true in nature, it is by construction the essence of information technology, and, as we will further see, this should and would (given technological determinism) ultimately be the blueprint for society.

The convergence of technology and biology is tackled head on in a special issue of *Wired*, "Living Machines" (*Wired* editors 2004a). The opening article outlines the basic premises of that approach: "Scientific advances point to a startling conclusion: The nonliving world is very much alive" (Meyer 2004). This, according to the article, is true for all sorts of nonliving human creations ("markets and power grids have much in common with plants and animals"). But more important, this is true for network technology: "We're beginning to discern life processes at their fundamental level, and as we re-create these processes in silico, we're starting to see how they work in inorganic settings. It turns out that many of life's properties—emergence, self-organization, reproduction, coevolution—show up in systems generally regarded as nonliving" (Meyer 2004). "Computer programs," the author says, "procreate, too. Genetic algorithms mimic biology's capacity for innovation through genetic recombination and replication, shuffling 1s and 0s the way nature does DNA's Gs, Ts, As, and Cs, then reproducing the best code for further recombination" (Meyer 2004). With these properties now being built into network technology "the line between organisms and machines is beginning to blur" (Meyer 2004). The author gives an example of unmanned aerial vehicles (UAVs). Programmed with simple rules, the UAVs are able "to direct themselves better than any dispatcher could" (Meyer 2004). The UAVs are networked and self-taught: "If one is shot down over Afghanistan, all drones everywhere gain improved responses to that form of attack." Not surprisingly, this technological sophistication has been known in nature for ages. The way UAVs work "is precisely how bacteria develop resistance to antibiotics, only faster" (Meyer 2004). So far the metaphor is restricted to nature and technology. To conclude the trinity, another link is added—the social realm. "Oddly enough," he says, "our growing knowledge of life processes could have its biggest impact in the social sciences. Social systems, after all, are made up of interacting agents, i.e., people. When we become adept at applying these insights to the social sphere, we'll be able to run simulations that reveal, say, the conditions under which Iraq would reconstruct itself. At that point, the new science of life will help us not only live better, but live better together" (Meyer 2004).

What the digital discourse suggests is that social affairs are guided by a logic similar to that which governs systems of information and communication in both nature and technology. Alas, this similarity is not simply metaphoric and suggestive (as, for example, in the case of complexity theory. See, for example, Urry 2003). The argument is much stronger here, suggesting that social dynamics should follow those of nature. Moreover, it offers the practical medium through which such change is to take place—network technology. Network technology is seen as a medium between nature and society, facilitating social change in accordance with the laws of nature. The special feature on "Living Machines" ends with a call for arms—"Evolve" (*Wired* editors 2004b)—that sums it all up in a big crescendo and urges society to keep up with technology and nature: "A lot can happen in a billion years, but look closely and you'll see the same dynamic at work in every system, and at every scale." Whether it's "microbiology or geopolitics," the underlying dynamics are the same (*Wired* editors 2004b).

The direction of argumentation may change: technology might be seen as following the logic of nature or vice versa. In an essay entitled "The Computer at Nature's Core" (Channell 2004), the author turns the table to suggest that rather than technology being like nature, scientists are now gaining insights about nature from technology. Scientists increasingly "explain natural phenomena by invoking such man-made artifacts as the computer" (Channell 2004). Like information technology, the natural world "is the result of a series of yes-or-no choices that take place at the level of quantum mechanics" (Channell 2004). Hence, to understand how nature works, one needs to carefully examine computers. And indeed, "research into quantum computers has implied that matter itself processes information . . . This has led some in the pure research world to the controversial claim that the universe itself is governed by the laws of computation and is, in fact, a computer . . . the field of systems biology is explicitly predicated on a computational model" (Channell 2004).

One has to wonder about the tautological nature of the argument: technology is like nature, which is like technology. But of course it's not the truth-value of these arguments that concerns us here but their social meaning. Both versions of the argument convey essentially the same meaning: there is something natural about network technology; it is not simply another machine, another tool invented by humans, but an arrival at the very core of the laws of nature. So much so that nature itself recedes and is acknowledged to be nothing more than a giant computer.[2]

It is crucial to distinguish radical posthumanist approaches to technology, such as Michel Callon and Bruno Latour's Actor-Network Theory, or Donna Harraway's notion of the cyborg, to the digital discourse's conflation of technology and nature. Posthumanism does away with the discursive dichotomy of nature and technology. The cosmology of the digital discourse is of another order: it asserts the conflation of nature and culture in reality, but it maintains the discursive power of the distinction. It claims that with network technology (i.e., at this particular historical moment) *technology* has become more *natural*. In other words, while posthumanist thinking undermines our notion of the nature-culture dichotomy, the digital discourse in fact reaffirms and celebrates this dichotomy.

INDUSTRIAL TECHNOLOGY VERSUS NETWORK TECHNOLOGY

The digital discourse is manifestly reflexive of, and responsive to, the general critique of technology, particularly the strain of dystopian critique that views technology as dehumanizing or as coming between humans and their nature, a critique most famously crystallized by Martin Heidegger (1977) and Jacques Ellul (1964; see Borgmann 1984a; Cooper 2002; Murphie and Potts 2003). In this context, and with the narrative of naturalization, the digital discourse insists on a break *within* the history of technology. Both Heidegger and Ellul write within a specific techno-historical context. Heidegger laments the rise of industrial technology at the turn of the twentieth century; Ellul fears large-scale technological systems enabled by the first wave of information technology in the 1950s, such as energy production and distribution. By evoking the radical break that is network technology, the digital discourse is able to incorporate and respond to this type of critique of technological utopianism and, by extension, respond to the critique of instrumental rationality in general.

It is not technological utopianism *per se* that the digital discourse advocates but a new utopianism based on a new technological paradigm.

Moreover, even within the technological paradigm of Toffler's Third Wave there are transformations so that not all information technology is created equal. I have pointed out already in Chapter 5 the shift in emphasis from *information* and *communication* toward the *network* as the central locus of contemporary technological revolution. Another fault line was drawn in the mid-1980s between the promise of big, centralized, mainframe computers (which have failed and gave us technocracy) and the personal computer (PC), which was construed as the carrier of a new promise of liberation (Dean 2002, 79–113). While both were in principle the same information technology, the PC was construed as the *opposite* of the supercomputer and, in fact, as a response to the pitfalls of the centralized computer. The latter accounted for an increasingly centralized, hierarchical, controlled, and technocratic society; the former promised to deliver a more decentralized, dehierarchized, and freer society.

The narrative regarding the naturalization of technology allows the digital discourse to reject a naïve, unreflexive version of technological determinism and technophilia. Those have been abundant in modern Western culture, perhaps even more so in American culture (Nye 1994; Segal 1985). As a general stance technological utopianism has been criticized through and through. Machines yielded greater productivity, yet they did not bring about the emergence of a "leisure society"; food processors chopped vegetables faster than women, but they did not necessarily liberated them from the toils of house labor (Cowan 1985; Wajcman 2004). How in posteuphoric times can the digital discourse still put so much trust in technology? The answer has a lot to with the naturalization of network technology. The construction of this narrative projects the message that the digital discourse is not duped by technology *per se* but rather specifically enthuses about *network* technology. Hence the recurring distinction in the digital discourse between network technology and industrial technology, specifically in regards to the measure of how natural these two technological paradigms are. In this formulation, industrial technology is constructed as the "Other" of network technology, upholding the positive "transforming fire" (Kelly 1998, 1) of digital technology *vis-à-vis* the negative effects of industrial technology and affirming the digital discourse's optimism in network technology.

It is in this context that the digital discourse insists on a fundamental break between industrial, mechanic, analogue technology and the new network technology. Yes, concedes the digital discourse, with all the benefits of Second-Wave technologies, in the language of Alvin Toffler, its utopian promises have failed, and its damages were overlooked.[3] But Third-Wave technologies are a new story, and utopian dreams regarding their emancipatory and progressive potential are right to be rekindled. This mindset has already been quite ingrained into our common sense and language, where network technology is habitually referred to as "high"-technology, asserting its superiority *vis-à-vis* good old (but full of false promises) "modern," "heavy," or "industrial" technology.

The clear distinction between old and new technology is based in the digital discourse on the assumed compliance of the latter with the laws of nature. Nicholas Negroponte asserts that cyberspace and information networks "look and feel . . . much more biological, taking [their] character more from flora and fauna than from the unnaturally straight-line geometry in artifacts of human design" (Negroponte 1997). Negroponte is able to distinguish between two types of technology—natural and unnatural—and by that he achieves two goals: first, naturalizing and reifying network technology and, second, fending off a generalized critique of technology; technology might not have been emancipatory and benevolent in the past, but that is only because it has been unnatural. That has now changed with network technology.

THEOLOGIZING TECHNOLOGY:
TECHNOLOGY AS TRANSCENDENCE

The rhetorical association of technology to nature evokes not only "truth," "reality," and "inevitability" but also appeals to our sense of transcendence as well and to the perception of nature as a secular sublime (Nye 1994). Technology, according to the digital discourse, is the locus of transcendence. By transcendence I mean here not only the overcoming of social conflicts and human ordeals by technological means (this type of "profane transcendence" was discussed in previous chapters) but also a more metaphysical experience of technology as elevating humans and the social to a realm beyond the mundane. According to the digital discourse, information, communication, and computing are not only the bases of contemporary society; they are metaphysical and transcendental as well.

AND GOD CREATED COMPUTERS

In a special issue of *Wired* about the intersection of science and religion, Kevin Kelly proposes that the universe, God, and its creation are all a computer. The title of the article—"God Is the Machine"—puts this assertion quite bluntly. Evoking the opening verse of the bible, the subtitle reads, "In the beginning there was 0. And then there was 1." Thus

Kelly reflects "on the transcendent power of digital computation" (Kelly 2002). A few of the naturalizing themes of the digital discourse are reiterated in this article as well. But they are now set in a different, transcendental context, and so their meaning is also changed. Life, Kelly says, is made of numbers: "Biology, that pulsating mass of plant and animal flesh, is conceived by science today as an information process . . . hard matter is information as well . . . [The physical world is] incorporeal. . . . everything . . . in the universe [is] . . . made of nothing more but 1s and 0s. The physical world itself is digital . . . Every *it*—every particle, every field of force . . . —derives its function, its meaning, its very existence entirely from binary choices, *bits*. What we call reality arises in the last analysis from the posing of yes/no questions" (Kelly 2002).

We have seen that type of assertions above. But Kelly now takes this "world-as-information" argument a little further, arguing for a new science, which Kelly terms "digitalism," which suggests that "the universe itself is the ultimate computer—actually the only computer" (Kelly 2002). Digitalism's attempt to understand physics as "a form of computation" opens the door, according to Kelly, to a new cosmology, a new understanding of nature and being, and ultimately a new theology: "From this perspective [that is, digitalism], computation seems almost a theological process. It takes as its fodder the primeval choice between yes and no, the fundamental state of 1 or 0. After stripping away all externalities, all material embellishments, what remains is the purest state of existence: here/not here. Am/not am. In the Old Testament, when Moses asks the Creator, 'Who are you?' the being says, in effect, 'Am'. One bit. One almighty bit. Yes. One. Exist. It is the simplest statement possible" (Kelly 2002).

Binarism, or digitalism, that is, the expression of reality as strands of 0s and 1s (or more accurately, as a state of existence or inexistence), is the ultimate model of reality (which in effect means it is not a *model* at all but reality as such), a universal language. Hence, it is divine. God, which knows reality for what it is, knows digitalism. Kelly seems to find in the Bible confirmation for digitalism. God's reply to Moses evidently vindicates digitalism as the ultimate reality. (And incidentally reveals the creator to be a programmer, or at least well conversed in the digital language.) But one should not really be surprised—after all, if God has created this world, he *had* to be a programmer. To be God is—quite literally—to play a programmer. Here is how the work of God-*cum*-programmer is presented by Kelly: "All creation, from this perch, is made from this irreducible foundation [0s and 1s]. Every mountain, every star, the smallest salamander or woodland tick, each thought in our mind, each flight of a ball is but a web of elemental yes/nos woven together" (Kelly 2002).

The power of computing, Kelly suggests, arises from its transcendental, transhistorical universality. He posits two assumptions: "all computation can describe all things," that is, anything from logical arguments to emotions can be translated into a digital language, and "all things can compute," that is, not only what we usually refer to as computers but also everything, from the human brain to metals and stones, perform computations. A third postulate, derived from the first two, is that "all computation is one," that is, that "all computation is equivalent" (a postulate also known as "universal computing"). Indeed, computing pioneers such as Alan Turing and John von Neumann extended "the laws of computation away from math proofs and into the natural world," declaring that "evolution and learning . . . were types of computation. Nature computed." Refusing to limit himself to the parochialism of planet Earth, Kelly ponders about extending these insights into the universe ("If nature computed, why not the entire universe?") and, even further, into metaphysics ("God, or at least the universe, might be the ultimate large-scale computer"). Computers, according to Kelly, are ultimately not human inventions but a divine creation: "In a sense, nature has been continually computing the 'next state' of the universe for billions of years; all we have to do . . . is 'hitch a ride' on this huge, ongoing Great Computations" (Kelly 2002).

It is critical once again to insist that the notions of the universe as a computer and nature, life, and existence as computing processes are not simply metaphors. In fact, the power of this rhetoric lies precisely in conflating the model with reality, since such *knowledge* presents itself as *truth* (Aronowitz 1989). Such an argument is not therefore "the-universe-as-computer"; the argument instead is that the universe *is* a computer. Kelly clarifies the distinction by asking one scientist to comment on the view of another: "[Do you] go along with the weak version of the ultimate computer, the metaphorical one, that says the universe only seems *like* a computer? Or do [you] . . . embrace Fredkin's strong version, that the universe *is* a billion-year-old computer and we are the killer app[lication]?" The scientist responds, "I regard the two statements as equivalent . . . If the universe in all ways acts as if it was a computer, then what meaning could there be in saying that it is not a computer?"—a rhetorical question to which Kelly answers, "Only hubris" (Kelly 2002).

The idea of the universe as a computer is presented quite literally by Kelly, who attempts to quantify the computing power of the universe. He reports on research published in the *Physical Review Letters*, in which an MIT professor calculated the computing power of the universe, or more accurately, "the upper limit of how much computing power the entire universe (as we know it) has contained since the beginning of time." The professor then compared this number to the computing power humans have built: "He then tallied up the total energy-matter available in the known universe and divided that by the total energy-matter of human computers expanding at the rate of Moore's law." His conclusions led him to predict that humans need some six hundred years "before all available energy in the universe is taken up in computing. Of course, if one takes the perspective that the universe is already essentially performing a computation, then we don't have to wait at all. In this case, we may just have to wait 600 years until the universe is running Windows or Linux" (Kelly 2002). And other scientists concur: "there is nothing theoretical to stop the expansion of computers"; "In the end, the whole of space and its contents will be the computer. The universe will in the end consist, literally, of intelligent thought processes" (Kelly 2002).

Taking this story back to the starting point of religion, Kelly reminds us that the equation of information with reality, the universe, and God is not new:

> Central to the evolution of Western civilization from its Hellenistic roots has been the notion of logic, abstraction, and disembodied information. The saintly Christian guru John writes from Greece in the first century: "in the beginning was the Word, and the Word was with God, and Word was God." Charles Babbage, credited with constructing the first computer in 1832, saw the world as one gigantic instantiation of a calculating machine, hammered out of brass by God. He argued that in this heavenly computer universe, miracles were accomplished by divinely altering the rules of computation. Even miracles were logical bits, manipulated by God. (Kelly 2002)

At this point it also becomes evident what significance Kelly finds in theologizing technology and in bringing religion back to the picture. The notion of universal computing—this most basic rule of the universe, or the "operating system" of the cosmos—is, according to Kelly, what binds us all together. In this "mystical doctrine of universal computing . . . we are linked to one another, all beings alive and inert, because we share, as John Wheeler said, 'at the bottom—at a very deep bottom, in most instances—an immaterial source.' This commonality, spoken of by mystics of many beliefs in different terms, also has a scientific name: computation" (Kelly 2002).

Information technology binds us together not simply through the banal and ubiquitous tools of communication, such as e-mails or blogs, but also through its evocation of the most basic reality of the universe and life. Notwithstanding material differences in the world, at the bottom we (humanity, Earth, and the universe) are all one. Kelly concludes his article by quoting Danny Hillis (a key figure in the development of computing) saying, "[Computation] has an almost mystical character because it seems to have some deep relationship to the underlying order of the universe" (Kelly 2002).

The illustration for Kelly's article captures the essence of the transcendental undertones of the digital discourse. At first glance one sees a typical glass vitrage commonly used to adorn churches. Indeed, it is a staple of religious artifacts. On a closer examination, the content of the glasswork seems to suggest a new variation in accordance with digitalism on our established theme of religion. Three elements are worth noticing. The centerpiece of the glass features a computer processor: three dimensional, golden, and most significantly "glowing." Is that the new, digital version of the golden calf, a sacred object of transcendence?

Another element in the glasswork is the "holy scripture" on top of the computer processor, which simply reads "1001000011010100." Whatever that might stand for, it most significantly stands for the signs themselves: God, and its creation, is digital, and the holiest of holies is simply a binary code. The third element graphically reiterates Kelly's assertion of digitalism as a unifying religious concept for humanity. The bluish rounded pieces of glass that circle the sacred processor feature the symbols of major Western and Eastern religions: Christianity, Islam, Judaism, Hinduism, Paganism, Taoism, and others. What this seems to suggest is that what Kelly terms Digitalism is a form of metareligion, which is able to bring all religions under its wing, since it reaches at the very essence of transcendence and metaphysics.

REENCHANTING INSTRUMENTALITY

The cover of the special issue about science and religion (2002) also deserves our close attention (Figure 7.1). The overarching element of design is a religious artifact, the cross. At the center of the cross is a human being—a man, in fact—small and fragile. However, he is surrounded and empowered by augmentation devices of science and technology. Among them are a computer processor, a microscope, artificial (and larger-than-life) arms and hands, a satellite, a human embryo, an apple (which might stand for genetic engineering), and so forth. Extending from the human's legs is the DNA double helix—an informational code from which, according to the digital discourse, human beings emerge. Albeit now, with information technology, this informational code denotes not determinism but opportunity. It is no longer divine providence programming the creation, but humans taking part in the process of re-creation. As biophysicist Gregory Stock says in an interview on cloning, "we are becoming the objects of

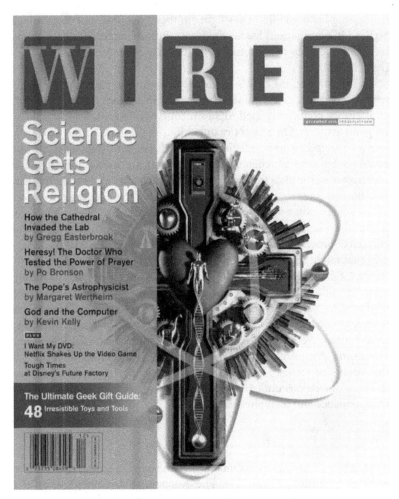

Figure 7.1.   *Wired to God?* Cover, *Wired*, December 2002 (reprinted by permissions from *Wired*/© Condé Nast Publications Inc.)

our own technological processes. We are seizing control of our own evolution" (Simpson 1997). Humans have turned from passive "consumers" of godly plans to active participants in the production of their destiny; what Toffler has termed "prosumers" (Toffler 1980, 265–88).

While the premise of the illustration is the blending of science and religion, it is significant that it is science that is embedded into religion rather than the other way around. It is the classic image of neurons circling an atom's nucleus that surrounds the cross and not vice versa. What is important for our discussion here is, of course, the mixture of the rational with the irrational, the mind with the heart, the material with transcendence, instrumentality with meaning. Because at the heart of the transcendental narrative is a reaction to the critique of modernity, rationality, and technology from Max Weber, through Jacque Ellul and Martin Heidegger, to Hebert Marcuse and Theodor Adorno, all of whom seem to have decried the disenchantment of the world with rationalization. The discontents of modernity, this narrative suggests, might have been the reality of previous modernities and therefore previous technological paradigms. What the digital paradigm does, according to the digital discourse, is overcome the chasm represented by these dichotomies and bring those dualisms into a harmonious unity.

At the center of the cross in the cover's illustration is a heart. But this time, and in contrast to the other natural objects in the illustration, this is not the biological heart and does not stand for advances in heart surgery, or artificial implants; instead it is the heart as a symbol of love, emotions, intuitions, and beliefs. Instrumental rationality, science, and technology, the image suggests, are not antonyms to meaning and transcendence but sit very well with them. The dualism is overcome to such a degree that, in effect, the opposing forces become one and the same thing: means become indistinguishable from ends, instrumentality becomes indistinguishable from meaning, and technology becomes indistinguishable from religion and transcendence.

The front cover of the August 2005 issue of *Wired*, which celebrates the World Wide Web's tenth anniversary, features another representation of the divine power of network technology.[4] The cover illustration makes a reference to Leonardo Da Vinci's famous *Creation of Adam*, one of the most recognizable iconographies of the biblical creation story. Part of the Sistine Chapel, *Creation of Adam* depicts God's hand reaching out to touch, and in effect animate, the first human. In the rendition on *Wired*'s cover, divinity reaches out and animates the hand icon commonly used to indicate hyperlink in Web surfing. Did God invent the Web? Is network technology divine and transcendental? *Wired* answers with a resounding "Yes."

The significance of the theologizing of technology is further illuminated by its relationship with the narrative regarding nature. Looking back at the history of American technological narratives, David Nye (2003) concludes that "in the American beginning, after 1776, when the former colonies reimagined themselves as a self-created community, technologies were woven into national narratives." In this American story of origin, "America [was] conceived as a second creation built in harmony with God's first creation," that is, nature (Nye 2003, 1).

TELEOLOGIZING TECHNOLOGY: TECHNOLOGY AND HISTORY

One theme we encountered in the discursive convergence of technology and nature is the idea that technology is autonomous from society, that is, that it has its own history independent of social history; what is more, in some sense the history of network technology precedes social history, both chronologically and ontologically. In such a framework, technology is reified and in turn conceived as an environment (like nature) to which humans need to adapt (Feenberg 1991). Moreover, such framework provides a cosmology that is not human-centric. Just as humanists, or natural historians, put humans and nature, respectively, at the center of their investigation, so does the digital discourse's technologism construct a history of the world in which technology is the main protagonist, the central axis of analysis.

MARGINALIZING HUMANS IN TECHNOLOGICAL HISTORY

Put differently, just as network technology is distinguished from industrial technology, so does each have its fitting cosmology. Since industry was "artificial" and man-made, a human-centric cosmology was fitting. But with informationalism, since network technology is "natural," the role and status of humans in that cosmology recedes. In such a history, technology has its own teleology (distinct and autonomous from human history) at the end point of which technology, rather than humanity, rules supreme. Human civilizations are therefore seen as transitory phases in a greater history in which humans are likely either to disappear, leaving the scene to technology, or to integrate into technology.

A forecast about the future of humanity is laid out, for example, by Vernor Vinge, academician and science fiction writer (Kelly 1995). His forecast incidentally reveals the digital discourse's assumptions regarding the secondary role of human history within a larger history: that of natural and technological information and communication. Vinge says, "If we ever succeed in making machines as smart as humans, then it's only a small leap to imagine that we would soon thereafter make—or cause to be made—machines that are even smarter than any human. And that's it. That's the end of the human era—the closest analogy would be the rise of the human race within the animal kingdom" (Kelly 1995). The significance of such a statement lies in turning our common sense of history on its head. In contrast to our modernist, human-centric history where humans are held constant, Vinge offers a story where technological progress is held constant and, as a consequence, humans assume a transient and ephemeral position. *As technology*, that is, as information processing machines, humans are quite limited. Put differently, the human is a quite primitive technology and is therefore (within the history of network technology) bound to vanish or be transcended by smarter machines.

This scenario, Vinge explains, is plausible since human agency *vis-à-vis* technology is increasingly receding, leaving technology ever more autonomous to follow its own course, or to assume full agency: "I think technology is close enough to being out of control that human intervention has become a weaker and weaker constraint." He concludes with a statement that puts together most of the technologistic assumptions regarding technology being autonomous and having its own history and teleology, a history and teleology that is linked with nature much more than with (human) culture: "It's important to regard technology in the long sweep of history as being one with history. In fact, it's one with biology, one with the rise of multicellular life forms, and it's headed someplace—probably. But it's not alien to the sweep of development and beauty and order in the universe" (Kelly 1995).

In such a techno-centric history the status of humans is revisited. They are no longer seen as the crown of creation nor the pinnacle of the natural world but are reevaluated *vis-à-vis* technology. In such framing they are portrayed as a dumbed-down version of computers. But this is not just a reevaluation of capabilities and performance or an acknowledgment of the limited computational capabilities of humans; it is rather a break with the core philosophical underpinning of humanism. This techno-centrism means that humans are no longer seen in the light of Kantian humanism as ends in and of themselves but in a new light, which might be termed superhumanism, or cyberhumanism (which should be distinguished from posthumanism), where human evolution is seen as a stage on the way to a higher end where more advanced levels of technological (and civilizational) developments are achieved. Humans are seen as a subdigital species in need of assimilating into a new species, which is increasingly more technological and less humane.

I must remind the reader that my task is not to engage with the argument philosophically or present moral or ontological critiques. Rather, I want to point out the cultural meaning of such a stance and its social and political ramifications in the context of contemporary capitalism. Set within this framework, such a narrative should be understood as depoliticizing technology, indeed deculturalizing technology, by reification and naturalization and by decentering the role of individuals and society in the creation of the conditions under which humans live.

TRANSCENDING HUMANISM

Another famous narrator of the superhumanist variant in the digital discourse is Hans Moravec. In his *Wired* portrait he is described as "a pioneer of robotics" and "our most gung-ho advocate of technology as a tool to transform human being and make us more than we are" (Platt 1995b). Why should humans be transformed, Moravec asks, and in what sense should we be made more than we are now? Moravec embraces posthumanism as a response to the limitations of humans. Compared with computers' calculating and rational capabilities, humans don't fair that well. The human form, Moravec says, "clearly isn't designed to be a scientist. Your mental capacity is extremely limited. You have to undergo all kinds of unnatural training to get your brain even half suited for this kind of work . . . You live just long enough to start figuring things out before your brain starts deteriorating. And then, you die . . . wouldn't it be great . . . if you could enhance your abilities via artificial intelligence, and extend your lifespan, and improve on the human condition?" (Platt 1995b).

Humans, according to Moravec, have been stretched to the limits of their capacity to handle human life, a life that is now spilling over into a new technological history. The world is "screwed up," Moravec says, because "we have a Stone Age brain, but we don't live in the Stone Age anymore. We were fitted by evolution to live in tribal villages of up to 200 relatives and friends, finding and hunting our food. We now live in cities of millions of strangers, supporting ourselves with unnatural tasks we have to be trained to accomplish, like animals who have been forced to learn circus tricks" (Platt 1995b).

While human capacities are limited, those of artificial intelligence, like a robot, can potentially grow infinitely. According to Moravec, when such robots reach a point where they learn and adapt (around the year 2020), they will become more like humans: "Success or failure will be defined by separate programs that will monitor the robot's actions and generate internal punishment and reward signals, which actually shape its character—what it likes to do and what it prefers not to do." Such robots will "mimic human traits." By 2030, those robots will "emulate higher-level thought processes such as planning and foresight." They will also be interlinked so they can learn from each other's experience, and they "will no longer need our help to show them how to do anything." Their interaction with humans will allow the robots to create their own consciousness. At this point, by 2040, "there will be no job that people can do better than robots," the robots will do everything, and we'll live in "a world of comfort, health, and boundless plenty" (Platt 1995b).

This utopian, end-of-history description might ironically turn out to be the end of human history as we know it. Moravec suggests a scenario in which some robots are able to break loose from the control of humans: "This marks the point where the genie finally gets out of the bottle," and humans find themselves faced with superintelligent machines. Under conditions of extreme competition between robots, Moravec does not expect humanity to last long. But "the takeover will be swift and painless" Moravec says. That's because "machine intelligence will be so far advanced, so incomprehensible to human beings, that we literally won't know what hit us." Robots will "find human activity interesting from a historical perspective," and they will remember us "as their ancestors"; According to Moravec, "We are their past, and they will be interested in us for the same reason that today we are interested in the origins of our own life on Earth" (Platt 1995b). In Moravec's scenario, humans are reenvisioned to remain part of the history of the world, that is, part of *technological* history, but this time not as prime subjects, but rather as objects. As Moravec puts it, "The robots will be a continuation of us, and they won't mean our extinction any more than a new generation of children spells the extinction of the previous generation of adults" (Platt 1995b).

To those who consider Moravec's forecasting a tragedy, he replies that "reversing the evolution of technology" would be even more disastrous. He therefore suggests reconciling with that future. We should "try to catch up with technology by accelerating our own evolution." As he puts it, "We can change ourselves . . . and we can also build new children who are properly suited for the new conditions. Robot children." Since technology has a rationale and a history of its own, since it responds to its own laws of evolution and since it is autopoetic, humans have nothing much to do other than to accept this technological reality. And besides, should his prognosis happens it will be "the ultimate form of human transcendence" (Platt 1995b).

CONCLUSION: NETWORK COSMOLOGY AND THE END OF HISTORY

The digital discourse's underlying theory of network technology and its place in society—its network cosmology—has been discussed by way of three central narratives:

- *The naturalization of network technology.* The digital discourse's theory suggests network technology is a natural construct, the technological condition's return to the order of the natural world, and an overcoming of the techno-human chasm. Contrary to the Industrial Revolution, the digital revolution is not so much an expansion of culture and rationality as it is a return to nature. This theory further implies a direct link between technological forms and social forms; technology, according to this theory, determines the shape of society.

- *The theologization of network technology.* Not unrelated to the claim of the return of technology to nature is a narrative that assigns network technology transcendental meanings. Network technology is not only progressive in the humanist sense but carries metaphysical and transcendental promises. It is therefore benevolent and right, indeed part of a "divine" plane.

- *The teleologization of network technology.* In the tradition of technological determinism, the development of network technology is seen as dissociated from human affairs. While the undertones of technology as nature are present here as well, this narrative offers a distinct voice stressing the asocial character of network technology. It suggests technology to be one with history—a motor of change and an end in and of itself—and implies the inevitability of its impact and the futility of any human and social intervention in its affairs.

According to the digital discourse, network technology constitutes the teleological climax of the history of technological progress, as well as that of information, the universe, and, indeed, being. The new realities of the network market, network work, network production, and the network human, which are in turn predicated on the realities of network technology—such as flexibility, adaptability, emergence, dehierarchization, decentralization, self-regulation, individualization, communication, and spontaneous order—represent a technological (and, in turn, social) revolution in the original (and forgotten) sense of the word (Barney 2000): a *return* to the very essence of nature and the universe. This new, network condition stands in stark contrast to the previous industrial condition, which represented a breaking away from this essence.

The spirit of networks is not simply a discourse about the overcoming of the discontents of industrial technology and the industrial condition (alienation, centralization, environmental degradation, and so forth) by network technology. More than that, it is a discourse about our civilization, our condition, our society constituting the end of history. This staple of contemporary political culture in advanced capitalist societies—epitomized in the maxim, at once postideological and programmatic, "There is no alternative"—receives technological clothing in the form of network cosmology. Network cosmology provides a technological validation for the crisis of ideological critique in the political culture of capitalist societies, that is, the widespread belief that such critique is futile and that we are living in a postideological time. More than that, network cosmology locates network technology—and with it technological reason, particularity the conflation of means and end, or the colonization of lifeworld by system—at the heart of a new constellation of society.

To sum up, the digital discourse's network cosmology is technologistic. By reifying technology, by suggesting a direct and nondialectic transfer between technological forms and social forms in which technology determines the shape of society, by naturalizing information technology and suggesting that it is akin to the forces of nature, by insisting that technological advance is human progress and that it is therefore benevolent, and, finally, by suggesting network technology to be one with history—a motor of change and an end in and of itself—and implying the inevitability of its impact and the futility of any human and social intervention in its affairs, this theory of technology depoliticizes and neutralizes (i.e., technologizes) what are essentially socially determined and politically contested power relations.

CHAPTER 8

NETWORKS AS THE TECHNO-POLITICAL CULTURE OF POST-FORDISM

IN ONE OF THE DEFINITIVE WORKS ON THE NETWORK society, Manuel Castells evokes Max Weber's essay on the spirit of capitalism to suggest that the "spirit of informationalism" is the cultural bedrock against which the network enterprise and informational capitalism operate (Castells 1996, 195–200). The tremendous economic and social transformations entailed by the rise of the network society, Castells suggests, are perhaps the result of material factors—namely, the restructuring of capitalism since the 1970s and the concurrent development and deployment of information and communication technology—but they must be accompanied by the rise of a new ethos, a new discourse to provide a general, cognitive map to interpret social life. Castells raises the question as to what the nature of this spirit might be but falls short of providing a systematic answer. This book provides a preliminary answer by analyzing contemporary digital discourse and uncovering the spirit of contemporary capitalism.

At the turn of the twenty-first century, a new "spirit of networks" has been emerging in the most advanced capitalist societies. The spirit of networks revolves around connectivity, flexibility, cooperation, decentralization, dehierarchization, spontaneous ordering, creativity and play, ad hoc assemblages, and, most prominently of course, the very notion of networks. These paradigmatic terms have become a dominant toolkit with which much of the dynamics and characteristics of contemporary society—pertaining to the economy, culture, politics, and sociability—are understood, experienced, and constructed. At the center of the spirit of networks—or rather, what holds the discourse together and binds its disparate themes—is network technology.

The purpose of this book has been to register and interpret this new technologically centered discourse, uncover its ideological and hegemonic dimensions, and tie this new discourse to the emerging social constellation of which it is a part. This concluding chapter offers two discussions. It first answers the empirical question, *What is the spirit of networks?* by reiterating the main empirical findings of the book. It then answers the theoretical question, *What is the sociopolitical meaning of the spirit of networks?* by locating this discourse within the transformation from a Fordist to post-Fordist society.

WHAT IS THE SPIRIT OF NETWORKS?

The metanarratives of the spirit of networks (flexibility, decentralization, distribution, and so forth) find their particular articulation in the narratives pertaining to the network market, network work, network production, and the network human, which are all anchored in a worldview of network cosmology. The spirit of networks extols the enhanced capacity of markets to be *self-regulated* and *self-governed*. Networks are characterized by the ability to foster *spontaneous order*, an order (rational and benevolent) that emerges from *bottom up* without any centralized planning, external governance, or oversight. Power in the network is located not with any group or structure; rather it is characterized as *distributed power*. Its particles can therefore be *assembled* by *individual nodes* in the form of *smart mobs* in order to bring about a desired outcome.

Network architecture is conducive to bringing about a new form of rationality based not on the top-down projection of abstract theories or grand-narratives but rather on a *swarm intelligence* based on the *wisdom of crowds* and the power of the network to render *dumb power* into rational results. By facilitating new, bottom-up constellations of action based on *disparate nodes*, networks contribute to *individual empowerment* by allowing individuals to come together in meaningful action. The spirit of networks, therefore, asserts the greater rationality of networks twice: as means and ends. Not only are networks yielding superior results (more profit and better products), but they also deliver more empowerment and liberation to the individual in the process.

Networks are in a constant state of *chaos*. The lack of centralized planning and oversight yields a system that is bent, and in fact thrives, on *perpetual flux*. Its dynamics are inherently *unstable* and *unpredictable*, and its trajectories are *nonlinear*. This fluctuating movement makes the network more *dynamic*, allowing for more openings and more opportunities—for business, culture, and so on—to come about. The architecture of networks allows for the *frictionless* and *streamlined* flow of substance and *interaction* between *disparate nodes* relieved from the hindrance of top-down regulatory obstacles.

To survive and thrive in the ecology of networks, nodes must face the increasingly *accelerated change* of their realities with the highest measure of *adaptability*. They must do away with any claim for stability, anchors, roots, and constants and substitute them with *flexibility*. In the network, there are no already established groups and structures based on essentialized categories and actors. Since networks are comprised of *unstable, nonessentialized nodes,* nodes, too, are characterized by a high degree of *flexibility*. Groupings in the network are *project-specific, ad hoc assemblages* of nodes. In such assemblages, nodes assume no essentialist homology and no long-term commitment to their groupings. This lack of commitment is reciprocated from the direction of the ad hoc project as well.

Networks offer a more *apolitical* mechanism for social interaction by embedding the dynamic characteristic of markets into the very workings of technology; the *hidden order* of spontaneous order, self-regulation, and the invisible hand are guaranteed by the very operation of network technology. Networks are therefore freer from political considerations and the oppressive powers of centralized institutions.

But networks do not simply make the operation of markets more perfect, or more akin to the market's ideal type; they also completely transform the activities entailed in the reproduction of society. Networks make the workplace *decentralized* and *dehierarchized*. With the *de-differentiation* between various echelons (workers, managers, and owners) the workplace becomes *flat*. Power relations are determined based on the criteria of *professionalism* and *meritocracy*, not on the ownership of the means of production or lack thereof. They are therefore not fixed and pre-established but ad hoc and changing according to the particular task at hand and the particular skills required. The multiple actors that once populated the workplace (the boss, the worker, the manager, and the CEO) are substituted by a single actor in the network workplace, which is defined first and foremost by its relations with network technology: the *digerati*.

The digerati construct their identity around the axis of network technology. They are characterized by being *technologically savvy, nerds* and *geeks*; being able to *multitask*, flexibly shifting between different projects and deploying different skills; and being *entrepreneurial*—not a cog in the system, but a node among competing nodes in the network. Even the most powerful of the digerati—its entrepreneurial tycoons—are essentially the same: they are passionate about *technological progress* and they deploy their technological expertise, not their financial prowess, to command power. The space of network work is flat and adverse to command, bureaucracy, and hierarchy to such an extent that these entrepreneurs themselves are bent on being *antiestablishment, negative, contrarian,* and *critical.*

Networks create a space that facilitates the positive blurring of work and leisure, of productive time and free time, of office space and home space, of public and private activities. Network work is ever present and can be performed everywhere. With networks, *play, joy,* and *passion* are reintegrated into the reproduction of society; the civilizational demands of labor are eroticized, and the chasm between those demands (for productivity *and* self-fulfillment) is overcome. Network work does away with bureaucratic, stifling organization and creates an airy, *debureaucratized,* and *decompartmentalized* structure of work.

The spirit of networks also heralds the coming of a new mode of production, which almost completely does away with the existing structure, experience, and meaning of work. Networks create a productive space that is, in fact, defined not by "corporations," "capital," "labor," and so forth but by the deconstruction of these structures and their reconstruction as homologous units of *prosumption* and *entrepreneurship.* Networks allow for each of their nodes easy access for *participating* in production. The spirit of networks heralds the demise of workers and in their stead come the *prosumers*—atomistic nodes of production and consumption that compete and *collaborate* with other production-consumption nodes (such as corporations) as *entrepreneurs.* All prosumption nodes are ontologically homologous.

Networks allow for the *dealienation* and *reenchantment* of the reproduction of society. *Creativity, self-expression, personal idiosyncrasies, passion, hobbies, individualism,* and *particularism*—all of these become not simply legitimate components of the productive process but become, in fact, assets to be harnessed into wealth creation. Network production is conducive to personal characteristics and, in that sense, allows for more *individual freedom, emancipation,* and *authenticity.* Networks are a space of *self-realization* through production; they therefore entail the *dealienation* of the productive process. But production and work become dealienated in another way as well: networks allow individuals to encounter products not simply as producers (whose activities are prescribed and who see only a fragment of the product) and not only as passive consumer, but as prosumers, a role that entails *control* and *deep engagement* with production and consumption, epitomized by the notion of *interactivity* and techniques such as *mass customization* and *social production.*

The spirit of networks is the spirit of perpetual *nonconformism* and paradigmatic *antiestablishmentarianism.* It is inherently *contrarian, countercultural,* and *counterhegemonic.* It holds deep disbelief in any grand project and *disloyalty* to any stable structures, such as "the company." Instead, the spirit of networks upholds loyalty to the very project of the network—fleeting assemblages and ad hoc crystallizations focused on a particular outcome—and, of course, to the very notion of network technology as progressive and benevolent.

For the spirit of networks, time and space have collapsed under their own weight. The spirit upholds an ethic of *temporality* (action over limited duration), *simultaneity* (multiple actions at the same time), *immediacy* (undelayed action), *fragmentality* or *nonlinearity* (action taking place over nonconsecutive segments of time), and *acceleration* (same action in less time) as positive new constructs of time. Likewise, regarding space it upholds an ethic of *nonspatiality* and *decentralization*.

The network—flat and dehierarchized, comprised of homologous nodes rather than structures—is more *participatory, inclusive*, and *democratic*. The spirit of networks upholds an ethic of *open source*, the paradigmatic enabler that allows individual nodes a stake in the network. The inclusive tendencies of open source and the network in general also make the network inherently *populist*, empowering the little guys, the underprivileged, marginalized, and off centered. The ad hoc and fleeting assemblages that characterize the network, and the lack of long-term institutional stability makes *collaboration* between nodes the prime mode of action, replacing competition. At the same time that networks allow collaboration around a project, they also help maintain the distinct *identity* and practical *autonomy* of nodes; nodes in the network are not pressured to assimilate into established structures and do not run the risk of losing their autonomy in the process.

The *reerotization*, or *libidinalization*, of the productive process projects on the whole human experience as well. Networks resonate with humans' true essence and help reconstruct a more *authentic* human experience. Like networks, humans, too, are *informational, multiple*, and *fragmented*. Networks allow for a better *commensurability* and a deeper interaction between humans and technology. Networks, therefore, facilitate the augmentation of human potentialities and allow humans to be more authentic and uncover their true selves. Hence, through networks and through the *integration* of humans and technology and the construction of the *cyborg*, the engagement of humans with the world becomes more meaningful, allowing more degrees of freedom.

Networks free humans from their essentialization: the network proves, vindicates, and augments their *nonessentialism*. Humans become what humans do (specifically, what they do with network technology), and their *pragmatic* being becomes a defining axis for who they are. As they are integrated into networks, their selves, identities, and experiences are *distributed* among multiple nodes—Web sites, servers, applications, and so forth. Being part of the network, humans shift swiftly and easily between multiple tasks: they are *flexible* and can engage with reality based on only *superficial*, temporary, and fleeting encounters. The spirit of networks upholds humans as *informational*: they are coreless (or unessential), comprised of a network of *components*, which can be *disassembled, distributed*, and *reassembled* in order to create a new whole.

Last, but not least, the spirit of networks is the spirit of *network technology* itself, the bedrock on which the network civilization is founded. Network technology is *natural*, a return to the essential characteristics and the underlying infrastructure—the "operating system"—of the cosmos: information, communication, networks, spontaneous order, adaptability, smart mobs, and so forth. The spirit of networks, hence, is an articulation of this fundamental, natural truth; indeed, it is the gospel of the *transcendental* powers of network technology. Network technology should not be seen simply as a human creation; instead it is essentially an *asocial* phenomenon, an unfolding of a *teleology* of increased information and communication since the dawn of creation.

WHAT IS THE SOCIOPOLITICAL MEANING OF THE SPIRIT OF NETWORKS?

Technology discourse, as a key social and political construct, is not a new phenomenon; it has its roots at the dawn of modernity and specifically with the rationalization of science and technology and the emergence of capitalism and the modern state. The spirit of networks is but the latest in this lineage. The purpose of this book has been not only to register anew the old themes prevalent in the discourse on technology—utopianism, technological determinism, instrumentalization, and so forth—but also to explore its contemporary particulars.

In more theoretical terms, this book has been concerned with both a *revival* and an *updating* of Habermas's critique of "technological consciousness" and his framework of "technology as ideology." It employed Habermas's framework as a theoretical model in order to explore the legitimation function of technology discourse, specifically in the context of the prevailing political culture and the structural constellation of state and market. As such, this framework served as a theoretical entry point into the discourse on technology in the contemporary post-Fordist era. In that sense, the revival of Habermas's critical lens and its application to the changing circumstances of the contemporary era has also led to its updating. The generational transformations of the political, economic, and technological pillars of Western capitalist society that have taken place since Habermas laid out his theoretical formulation beg for such updating. This book was concerned with elucidating how technology discourse works under a particular social constellation, highlighting its contemporary hues.

Simply put, the answer to the question "what is the sociopolitical meaning of the spirit of networks?" is twofold. First, the spirit of networks is the discursive "facilitating shell" (Fraser 2003), the "cultural logic" (Jameson 1991), or the legitimation discourse of post-Fordist society. The spirit of networks is understood in this book not simply as a *description* of contemporary techno-capitalism but also as central to the *constitution* of the novel ways of life, institutional arrangements, and social power relations that it entails. The spirit of networks is central to how contemporary life is understood, experienced, and practiced. It is the connective tissue between two key transformations of the last four decades: the shift to a post-Fordist society and the rise of network technology. It is the discourse that legitimates and enables these transformations.

A New Trade-off between Alienation and Exploitation

Second, the spirit of networks has been constructed as a response to the humanist critique of Fordist society. It transcends the technological, institutional, and social pitfalls that prevailed during Fordism—which were the subject of the humanist critique—and therefore marks a watershed in the shift of legitimation function of technology discourse from Fordism to post-Fordism. During the Fordist phase of capitalism, technological discourse gave legitimation to a socioeconomic regime of accumulation that responded to a social critique of capitalism while it downplayed, and even ignored, the humanist critique. Industrial technology, the assembly line, the bureaucratic corporation, the statist regulation of the economy, and the provision of welfare were all conceived in the technology discourse during Fordism as technologies and techniques that respond to concerns put forth by the social critique, such as the need for social security, stability, and equality, that is, those geared toward mitigating exploitation.

The critique of Fordism, to which the spirit of networks responds, therefore, targets precisely the oppressive nature of the administered state and the bureaucratic corporation, the loss of personal authenticity, and the deeroticization of the productive process, that is, the harmful ramifications of Fordism in terms of alienation. Their amelioration has, in fact, become the centerpiece of the spirit of networks that emerges in the post-Fordist phase of capitalism. With post-Fordism, technological discourse has come to legitimate a new socioeconomic regime that responds to the humanist critique of capitalism while it downplays the social critique. Network technology, the lean and flexible corporation, and flexible modes of employment and production are all conceived in the discourse on network technology during post-Fordism as technologies that respond to concerns put forth by the humanist critique, such as the need for individual empowerment, authenticity, and creativity, that is, those geared toward mitigating alienation.

While industrial technology during Fordism was used to legitimize the social compact between capital, labor, and the state, network technology during post-Fordism legitimizes precisely the decomposition of this compact and the constitution of its alternative: privatized relations within the context of a global market and of civil society. While Fordism extolled the power of the old technology in the name of social equity and stability as a public and political project, post-Fordism extols the power of the new technology in the name of individual authenticity and liberation as a private and apolitical (or postpolitical) enterprise.

The new legitimation discourse of networks is woven along the axis of the critique of the techno-political culture of Fordism, that is, its paradigmatic technological forms and the organizational, institutional, occupational, social, cultural, political, and personal constellations that are assumed to stem from it. The responses to the pitfalls and shortcomings of the Fordist techno-political culture are now integrated into the new techno-political culture of post-Fordism. To put it in terms of the framework that had Daniel Bell worry at the turning point between those two phases of capitalism: with the spirit of networks, *The Cultural Contradictions of Capitalism* (1976) have now been resolved; the new capitalism does not reject, but quite the contrary—it harnesses self-realization, personal expression, and hedonistic impulses into its mode of operation.

In that context, network technology is constructed in contemporary technology discourse as amending the pitfalls of Fordist production by responding to concerns regarding individual emancipation and harnessing those human facets that have been suppressed during Fordism—individualism, authenticity, creativity, personal expression, and so forth—into the productive process. However, at the same time that network technology is constructed as responding to these individual demands, it is also construed as requiring the downplaying, and even rejection, of the concerns for social emancipation, the response for which has been epitomized by job tenure, employment security, embedded markets, and the provision of welfare and social security, which characterized Fordist society. The spirit of networks offers a new trade-off between the societal and the individual, between socially backed security and individual opportunity, between long-term stability and liberating flexibility; in short, between the social *safety net* that was the ideological underpinning of the Fordist, Keynesian, welfare state and the promise of the *Internet*, so to speak, for personal liberation and dealienation (Fisher 2007). The dynamics of the new capitalism powered by network technology—flexibility, adaptability, temporality, spontaneity, and so forth—are constructed in contemporary technology discourse as characteristics that both promise individual emancipation and negate the possibility for social emancipation.

To sum up, this book has argued that the emergence of a new capitalism and a new technology has been accompanied by the rise of a new technology discourse that offers a new type of legitimation to capitalism. During the Fordist phase of capitalism, technology discourse extolled the capacity of technology to enhance social goals of security, stability, and equality by mitigating the *exploitative* nature of capitalism. In contrast, during the contemporary, post-Fordist phase of capitalism, technology discourse extols the capacity of technology to enhance individual goals of personal empowerment, authenticity, and creativity by mitigating the *alienating* nature of capitalism. By moving from the mitigation of exploitation to the mitigation of alienation, I further suggest, contemporary technology discourse legitimates new constellations of power entailed by the new capitalism, at the heart of which is the weakening of labor and the state *vis-à-vis* capital, the liberalization of markets, the privatization of work, and the increased flexibility of employment.

THE SPIRIT OF NETWORK AND THE LEGITIMATION OF CAPITALISM

The legitimation function of the spirit of networks has been studied here primarily as it pertains to economic life (e.g., employment, the market, production, consumption) and as a general template of political culture. Since the legitimation scope of the spirit of networks pertaining to the particulars of economic life have been discussed extensively in the previous chapters, let me sum them up briefly and then move to a more abstract and general discussion regarding the significance of the spirit of networks as the techno-political culture of contemporary post-Fordist society.

The spirit of networks legitimizes the withdrawal of the state from the planning, management, and regulation of the economy and from its welfare obligations (Chapter 3); the move from a national protective economy to a globalized, deregulated, and unitary market; the privatization of work (Chapter 5); the eradication of "work" and "working class" as viable social categories (Chapter 4); the insulation of the economy from the democratic political process, or the disembeddedness of markets from society (Chapter 3); new employment schemes that are more precarious and flexible (Chapter 5); a new, decentralized disciplinary regime; a shift to a placid political discourse of a classless society, devoid of antagonism and contradictions (Chapter 4); a reorganization of working life and of labor processes, which are more flexible (Chapters 4, 5, and 6); the depolitization of economic affairs as technical, technological, and instrumental; and the construction of contemporary social constellations as postideological, representing the end of history (Chapter 7).

The spirit of networks does not simply describe the new realities of postFordism. It also legitimizes them by framing them within a technologistic framework, that is, by tracing them to their roots in network technology. By doing so, it depoliticizes, neutralizes, naturalizes, and universalizes social processes that are politically charged and particularistic. This is perhaps most readily evident in the keywords that dot the discourse and that are taken quite directly from the technical terminology of information networks professionals: "networks," "distribution," and "nodes." Simply put, the spirit of networks legitimizes the wholesale transformation entailed by the shift from the Fordist phase of capitalism to its post-Fordist phase. Table 8.1 summarizes these transformations along some key coordinates.

Table 8.1.   Key coordinates in the transformation from the Fordist to the post-Fordist phase of capitalism

	Fordism	Post-Fordism
Political culture	Social democracy	Neoliberalism
Organization of labor	Labor movement	Individual contracts
Scope	National	Global
Mode of production	Mass production	Mass customization, "just-in-time"
Economic organization	Big corporation Centralization	Lean production, outsourcing
Working arrangement	Tenure, advancement, career path	Flextime, temporary
Work process	Routine (same procedures)	Variable (multitasking)
Technology	Industrial/mechanical technology	Network technology
Technological Legitimation	Interventionist national state	Deregulated global market
Locus of political legitimation	Democratic state	Technological networks

What we see with the spirit of networks is a discourse that is able to respond to both the systemic concerns of capitalism and its humanist critique. The digital discourse is the legitimation discourse of a new phase of capitalism. It offers a spirit of capitalism that on the one hand sustains, cements, and legitimizes capitalism and on the other hand sets new tests for its legitimacy. The spirit of networks is still very much committed to the end of the old versions of capitalism inasmuch as it is ultimately about the endless accumulation of capital. But it is also a new spirit of capitalism to the extent that it provides new guidelines for the means of the new capitalism, demanding more participation, more democratization, more personal expression, and more interpersonal relations.

THE SPIRIT OF NETWORKS AND THE DECOMPOSITION OF THE FORDIST CLASS COMPACT

The spirit of networks should be understood not only in terms of the new particular and "local" demands of capitalism (such as the demand from workers to be flexible) but also in terms of a wholesale and "global" transformation in the political culture of post-Fordist society. At that level, the spirit of networks should be understood as a new legitimation model that responds to the crisis of the old legitimation model—that of the welfare state—and substitutes it (Habermas 1973). The spirit of networks facilitates the crystallization of a new constellation of power that characterizes the post-Fordist phase of capitalism. The tripartite social compact that characterized the Fordist, national phase of capitalism—labor-state-capital—has been flipped, as Ram (2006) puts it, substituting one actor—the state—with network technology as the third tip of the triangle. With network technology—and the social, economic, cultural, and political networks that it underlies—capital's route to labor (and vice versa) in post-Fordism is mediated by networks, thereby bypassing the state. The post-Fordist and global phase of capitalism is characterized by the constitution of (technological) networks as the containers, organizers, and mediators of the social relations between capital and labor (Ram 2007).

This new constellation is highly consequential for the question of legitimacy. By partaking in the social compact of Fordism, the state enjoyed a high level of legitimacy based on the power it could exert *vis-à-vis* capital and labor. In fact, the state enjoyed legitimacy to such a degree that it not only managed large parts of economic life but also played a central role in social and cultural life as well. Under post-Fordism, as the state is cut off from this tripartite compact, its legitimacy is hampered and in turn it loses its leverage as a dominant actor in social life, including in spheres pertaining to the management of material reproduction: work and the economy.

During Fordism, capitalist legitimation was mediated through the state, that is, through the technical and technological tools at the state's disposal: taxation, redistribution, centralized planning, fiscal and monetary control, and the provision of employment and welfare programs. In other words, technology played a role of support for state action; technology legitimized the centrality of the state in the reproduction of capitalist society. In post-Fordism, legitimation for capitalism has come to be mediated through network technology. Network technology provides legitimation not so much for an alternative global polity but for the ideology that now—with network technology—such polity is redundant. Such technological legitimation supports the depoliticization and desocialization of markets and the economy in general, or their dissociation from the political sphere.

One can therefore speak of two legitimation models, or two "triangles" of power. In the Fordist condition, the state enjoys political legitimation and exercises technical and technological administration of the economy. In that framework, technological legitimation is mediated by the state. In the post-Fordist constellation, the state loses its political legitimation for the administration of the economy, and instead technological legitimation is not mediated through concrete (and democratic) social institutions but through the internal operation of network technology, as encapsulated in the spirit of networks. In sum, with post-Fordism capitalist legitimation becomes more depoliticized and more instrumentalized; it becomes—quite literally—a *technological* legitimation rather than a form of political legitimation *through* technology. Hence, the discourse on network technology is part of a new social constellation where (network) technology is substituted for the political institution that is the state as a container of political legitimation.

TECHNO-POLITICAL CULTURES: THE CLOSURE OF THE POLITICAL

These sweeping transformations are detrimental to the scope and nature of the political as such in society. In that context, the discourse on technology has in both eras been an indispensable companion to the dominant political cultures of social democracy and neoliberalism, respectively. The techno-political culture of Fordism, and modernity in general, delivered planning, control, and stability; it rendered politics a project of reason, which leads to proactivity; it assigned politics great molding powers in society, and in that project technology played a central role.

In contrast, the techno-political culture of post-Fordism, and post-modernity in general, is reactive. It acknowledges its limited influence and legitimates a reality of minimum politics, and an adaptivist politics at that, upholding, in fact, the disengagement of politics from social affairs.

Here, too, technology is at the throne: network technology is constructed as a facilitator of self-regulation, delivering a flat playing field and democratic access without privilege—that is, a variety of mechanisms and dynamics that at one time needed a strong polity and now demand minimum, if not a complete insulation from, politics.

This brings us back to Jürgen Habermas and to the adaptation of his thesis to contemporary political culture. Habermas was particularly concerned about the closing down of the political and critical horizons in the administration of society; the collapsing of the lifeworld into the rationale of the system (Habermas 1984). Habermas's concern was raised—to use the conceptual framework of this book—at a time when the political culture was constructed as a response to the social critique of capitalism and, thus, was given the form of social democracy. The contemporary situation is not only different—founded on a neoliberal political culture, which is constructed around a response to the humanist critique—but also more troubling in terms of the political.

From the point of view of political critique, I argue, not all types of critique are created equal. There is a fundamental difference between the social critique and the humanist critique in terms of the *political space* that they allow, a difference that Boltanski and Chiapello (2005) ignore. Those two different types of critique create not simply two different political cultures (social democracy and neoliberalism), but rather they create political cultures that are qualitatively different precisely in the space they allow for meaningful political critique and action (for a critique on Boltasnki and Chiapello see Callinicos 2006).

This point has recently been elucidated by David Harvey (2005) in his rereading of Karl Polanyi's 1944 analysis of *The Great Transformation* (2001 edition). To a large extent, the demands for personal freedom—the core of the humanist critique—is in conflict with the demand for equality and solidarity—the core of the social critique (Harvey 2005). In the same vein, we can say that the response to the humanist critique, which dominates the spirit of networks, comes at the expense of responding to the social critique. Such analysis leaves us not only with the question of which critique is better—to which one can only respond with a normative answer—but also with the question of which critique is more political. This latter question can be given a theoretical response as well. A political culture founded on the social critique is inherently more political.

By its very definition and its very demands, the social critique opens up a political and critical space. The demands of the social critique are articulated in terms that are always-already society-wide, universalist, and political. Its demands for redistribution, equality, and welfare also require the embeddedness of the economy within political arrangements. The social critique cannot be responded to without strong political institutions and mechanisms. It is precisely this understanding that led the reformist (i.e., nonradical) strand of the working class to direct such enormous energies of legitimation in the direction of the state during Fordism. The response to the social critique during Fordism through a democratic process led to a political culture that is inherently more political where more aspects of society (particularly those concerning its reproduction) are embedded in the political process.

This stands in contrast to the political culture that characterizes a response to the humanist critique. The demands of the humanist critique are articulated in a language of individualism, personal autonomy, and the privatization of social life; it demands personal expression, authenticity, and individual engagement. While putting forth these demands is political in and of itself, in their effect and to the extent that they diminish a response to the social critique they result in what Marcuse described as "the closing of the political universe" (Marcuse 1991, 19). The response to the humanist critique during post-Fordism through market mechanisms leads to a political culture that is inherently more apolitical and where more facets of social reproduction are excluded from the political process, mediated, and resolved through technologically mediated networks. The closing of the political universe is coupled by an opening up and augmentation of the market universe and of a civil society conceived and fashioned after a market model.

Hence, while the spirit of networks is indeed constructed as a response to the critique of Fordist society, it does so in a manner that results in the depolitization of contemporary social relations: the spirit of networks resolves the *social* and *political* concerns that underlie Fordist society *technologically*. While closing off the political universe it opens a technological universe that reasserts the central role of technology in the resolution of social and political questions, in effect rendering these questions technological in nature.

The spirit of networks, with its explicit critique of the pitfalls of Fordism, legitimates the shift to post-Fordism first and foremost by depicting this shift as *progress*. Since the contemporary post-Fordist condition is intertwined with network technology—itself a higher development stage in the history of technology (see Chapter 7)—the social formations that stem from it are also conceptualized as being of a higher stage, a social development. The shift to post-Fordism, in other words, is understood as technologically induced and, therefore, as apolitical, asocial, and inherently progressive.

Moreover, the ideological thrust of the spirit of networks lies in inter-weaving this new trade-off between social emancipation and individual emancipation around the axis of network technology. The promise of the spirit of networks in terms of increased individual emancipation and dealienation is presented as *conditioned* by the reorganization of social practices and social relations in accordance with networks. Simply put, emancipation is constructed as demanding the further flexibilization and privatization of work.

Networks as Critique?

The discourse on networks in its broadest understanding has recently become the centerpiece not only in discussions on the economy, labor, globalization, and production but also in the discursive and analytical toolkit with which much of the contemporary world is understood, talked about, and ultimately constructed. Moreover, the idea that the discourse on networks is of significant critical potency has been making strides in diverse social milieus, from popular culture to political theory and phi-losophy. Thus, dehierarchization, decentralization, emergence, and distri-bution are seen as inherently critical to the established power structures, as undermining them, and, hence, as socially transformative, progressive, and even revolutionary. For example, much of the discourse and prac-tices of the so-called antiglobalization movement is founded on the spirit of networks. It entails the decentralized assemblages of disparate groups, which convene ad hoc around a specific political project. The movement is also dehierarchized: there is no overdetermination for any essentialist political analysis, project, or identity (Castells 1996). It is worth noting that to a large extent the target of this movement has been post-Fordist capitalism itself, and perhaps for that reason the spirit of networks is com-monly assumed to harbor a critical potential.

Likewise, recent social theory has also employed the discourse of net-works and elaborated on its critical potential. This is certainly true for academic and political discourses on cyberculture, hybridity, and cyber-feminism. But such approach also appears in theories concerned with critique of contemporary capitalism. Hardt and Negri (2001, 2005), for example, conceive of a decentralized, network-like "empire" and hope that the nonessentialized, atomized nodes of the "multitude" come together in meaningful political action (Hardt and Negri 2001, 2005). Others have offered a revival of the Italian autonomist tradition and note how the network structure of global capitalism might offer openings for a recon-stitution of labor struggles (Dyer-Witheford 1999). And Christian Fuchs (2008) theorizes that the Internet facilitates two dialectically antagonistic tendencies of competition and cooperation. The hegemonic competitive form of contemporary social relations thrives on the network structure of the Internet but also gives rise to an emerging cooperative potentiality.

These intellectual and political projects all attempt to elucidate and theorize a critical moment in the workings of contemporary, networked capitalist society. These are admirable projects, in my view, committed to one of the core goals of critical theory: that is, to locate the conditions of possibility for transcendence within the immanence. Likewise, both in theory and in practice, what critical potentialities are present in the spirit of networks should be extracted and exhausted in further research.

But what these positions fail to acknowledge is the degree to which the spirit of networks has been intertwined with technology and technological reason. The juxtaposition of the realities of contemporary capitalism and its spirit with the discourse on network technology—which has been a central purpose of this book—illuminates the technological blind spot in such analyses. Boltanski and Chiapello explain this lack as an attempt on their part to avoid the fallacy of technological determinism (Boltanski and Chiapello 2005, xviii–xix). To be sure, as Chapter 7 has shown, this fallacy is easy to slip into precisely because the new capitalism is so closely intertwined with network technology. The networked organization (Castells 1996), the command and control of headquarters in a global chain of production (Robins and Webster 1999; Harvey 1989), just-in-time production (Harvey 1989), the expansion in space of production, distribution, and consumption processes (Beniger 1986), the increased flexibility of the work process (Greenbaum 1995; Robins and Webster 1999)—network technology has been instrumental in bringing all these changes about, or at least in helping to materialize their full potential. Indeed the prominence of information technology in contemporary society has led many to call ours the "information society" (Bell 1999; Castells 1996, 1997, 1998; Duff 2000; Lash 2002; Mackay 2003; Mattelart 2002; May 2002, 2003; Robins and Webster 1999; Webster 2002; Stehr 2001).

However, by leaving network technology out of the realm of contemporary capitalism, such analyses overlook another dimension of the intersection of technology and society, which has been the centerpiece of this book: the role of technology as a legitimation discourse (Marcuse 1991; Habermas 1970), as the "religion" of an instrumental society (Noble 1999), and as the cultural fabric from which the contemporary spirit of capitalism is woven (Mosco 2004). Their analysis therefore fails to notice that the spirit of capitalism, both old and new, is inextricably linked with the discourse on network technology—the digital discourse.

The introduction of the digital discourse as a sociological object into the analysis of contemporary capitalist society and its understanding as a legitimation discourse point to the limited critical potential of the spirit of networks. The spirit of networks is too inextricably intertwined with instrumental rationality and systemic demands to offer a truly critical platform. As this book has shown, the spirit of networks is not a critical discourse, but precisely the opposite: it is the discourse of the new capitalism *par excellence*.

NOTES

CHAPTER 1

1. The various threads of the critical analysis of technology are commonly referred to as "science and technology studies" and include the social shaping of technology, the social construction of technology, and actor-network theory (see, for example, MacKenzie 1996; MacKenzie and Wajcman 1999; Law and Hassard 1999; Latour 2005; and Bijker, Hughes, and Pinch 1989).

CHAPTER 2

1. Since its inaugural issue, *Wired* has christened McLuhan as its "patron saint."

CHAPTER 3

1. The article later appeared in an extended book form with the same title (1998). References in the chapter refer to book pages.
2. Throughout this book emphasis is always in the original, unless specified otherwise.
3. By arguing that the digital discourse constructs individual nodes as irrational, I do not mean that the individual nodes are constructed as decidedly *anti*rational, but simply that they are devoid of system-wide, big-picture, theoretical, and abstract rationality.
4. It is worthwhile to point out the rhetorical tool used here in order to legitimize the idea of chaos. It is reminiscent of a joke about a borrowed kettle evoked by Freud (1963)—and recently retold by Žižek (2005)—to account for the nature of logic in dreams. In the joke, the kettle owner accuses his friend of returning the kettle damaged, an accusation to which the friend replies, "I have never borrowed your kettle; I retuned it to you unbroken; it was already broken when I borrowed it." In a similar fashion, Kelly suggests flux, chaos, and churning (along with their corollary social effects of instability and unpredictability) should not be opposed to or mitigated for three reasons: this cannot be done (flux in the network economy is inevitable; a transfer of a natural phenomenon into the social realm through network technology), it is better not to do it (flux is benevolent, yielding good results for everyone); and it is dangerous to do it (since it will result in knocking the system out of its self-regulating, natural imbalance). For these three, not necessarily compatible reasons, economic flux should be,

respectively, duly accepted, enthusiastically celebrated and encouraged, and not tempered with.

5. The term "neoliberalism" has come to occupy two meanings in common parlance. On the one hand it denotes an economic and political theory. On the other hand the term stands for the realities of contemporary capitalism, a political project. Harvey, for example, distinguishes between *utopian* and *political* neoliberalism (2005, 19). While those two constructs are highly correlated, they are not one and the same, and in fact at times the realities of neoliberalism as a political project conflict with its theory (Harvey 2005, 21). As a matter of clarity and simplicity and since this book is concerned with the realities of discourse, I limit my discussion here solely to the theoretical discourse of neoliberalism, not its political implementation.

6. In narrow disciplinary terms, Schumpeter is not part of the neoliberal school. But his ideas have been the intellectual bedrock of much of what is ultimately known today as neoliberalism.

7. It is important to emphasize that lessons and metaphors were drawn from the most mechanical aspects of the technology, how it is constructed, its minute components, their operation, and so forth, rather than any abstract or literary trope the clock might symbolize, such as "time" or "life."

CHAPTER 4

1. In this chapter, I am not using Wikipedia as an authoritative resource but as a way to tap into what concepts denote in the popular culture.

2. A search of the *Wired* archive yields a few hundreds references to the term "digerati" in the magazine.

3. See http://www.wired.com/wired/archive/11.06/office_spc.html. Most of the photos and illustrations referred to in the book can be accessed via the Web. A link is provided whenever that is that case.

4. Coincidentally, but noteworthy, the term "Generation X," which in Campbell's view became "generation equity," was popularized in 1991 by a book titled *Generation X: Tales for an Accelerated Culture* written by Douglas Coupland, author of *Microserfs*. Generation X is characterized by nihilism, cynicism, a postmodern attitude to life, a "so what" and "who cares" attitude, distrust on societal institutions, a "take the money and run" attitude toward work, childlessness, pessimism for the future, a *carpe diem* attitude toward living, and slacking (Wikipedia: Generation X).

CHAPTER 5

1. IPO stands for initial public offering: the first sale of a corporation's common shares to public investors.

2. An updated version of the older prosumer is the "produser," a hybrid of producer and user. The difference is mainly technical: the consumer of goods

of yesteryear has become the contemporary computer user who both uses and produces informational content.

3. See http://www.wired.com/wired/images.html?issue=13.08&topic=tech&img=1.
4. For example, the annual sales of these companies range from $14.8 billion (Eli Lilly) to $64.6 billion (Procter & Gamble). The cumulative annual net income of these four companies is $15 billion, and they employ 371,600 workers. Information on these companies was compiled from Forbes.com, Hoovers. com, and the Business and Company Resource Center, all of which are publicly available online.
5. See Mark Granovetter, "The Strength of Weak Ties," *American Journal of Sociology*, 78 (May 1973): 1360–80. Granovetter groundbreaking article made a theoretical connection between micro and macro analysis of the economy, arguing that it is weak individual ties (of acquantances) rather than strong ties (of friends) that have a greater role in the operation of the ecnomy.
6. Colgate-Palmolive's annual sales are $11.5 billion, its annual net income is $1.38 billion, and it has 35,800 employees.
7. http://www.innocentive.com.
8. See http://www.wired.com/wired/images.html?issue=14.06&topic=crowds& img=3.
9. *Online Etymology Dictionary*, s.v. "Company," http://www.etymonline.com.
10. *The American Heritage Illustrated Encyclopedic Dictionary* (Boston: Houghton Mifflin Company, 1987), s.v. "Company."
11. See Michael Perelman, *Railroading Economics: The Creation of the Free Market Mythology* (New York: Monthly Review Press, 2006).
12. The word "amateur" comes from the Latin root "love."
13. See http://www.wired.com/wired/images.html?issue=14.02&topic=lego&img=2.
14. See http://www.wired.com/wired/images.html?issue=14.02&topic=lego&img=5; http://www.wired.com/wired/images.html?issue=14.02&topic=lego&img=6; http://www.wired.com/wired/images.html?issue=14.02&topic=lego&img=7; http://www.wired.com/wired/images.html?issue=14.02&topic=lego&img=8.
15. See: http://www.wired.com/wired/images.html?issue=14.02&topic=lego&img=4.
16. The *Wired 40* is an annual list of 40 companies judged by the magazine to be "the most *Wired*." It is the digital equivalent of the *Fortune 100*, focusing not only on financial parameters, but also on parameters construed to embody the digital spirit, such as use of network technology and pursuit of innovation.

CHAPTER 6

1. In academic discourse, this idea is articulated in actor-network theory (Latour 2005; Law and Hassard 1999; Callon 1991).
2. This argument is developed in her book *Life on the Screen: Identity in the Age of the Internet* (1997).

CHAPTER 7

1. See, for example, Lang 2001; *Wired* editors 1998; Brand 1998; Sellers 1995; Dyson 1995; Westbury 1995; Schrage 1994.
2. See, for example, the illustration for the article by Channell (2004), available at http://www.wired.com/wired/archive/12.02/view.html?pg=2.
3. *Wired* in fact published articles criticizing past eras of technological utopianism. See for example, Jon Katz's "Lost World of the Future," *Wired*, October, 1995.
4. See http://www.wired.com/wired/coverbrowser/2005, issue 13.08.

References

Agger, Ben. 2004. *Speeding up fast capitalism: Cultures, jobs, families, schools, bodies.* Boulder: Paradigm Publishers.

Aglietta, Michel. 2001. *A theory of capitalist regulation: The US experience.* New York: Verso.

American Society of Magazine Editors. n.d. *National Magazine Awards Database of Past Winners and Finalists,* s.v. "Wired [Magazine Title]," http://www.magazine.org/asme/magazine_awards/searchable_database/index.aspx (accessed October 1, 2009).

Anderson, Chris. 2002. Spam-haters of the world unite! *Wired,* September.

———. 2006a. *The long tail: Why the future of business is selling less of more.* New York: Hyperion.

———. 2006b. People power. *Wired,* July.

Aronowitz, Stanley. 1989. *Science as power: Discourse and ideology in modern society.* Minneapolis: University of Minnesota Press.

———. 1994. Technology and the future of work. In *Culture on the Brink: Ideologies of Technology.* ed. Gretchen Bender and Timothy Druckrey, 15–29. Seattle: Bay Press.

———. 2001. *The last good job in America: Work and education in the new technoculture.* Lanham, MD: Rowman & Littlefield Publishers.

Aronowitz, Stanley, and William DiFazio. 1994. *The jobless future: Sci-tech and the dogma of work.* Minneapolis, MN: University of Minnesota Press.

Ashford, Nigel, and Stephen Davies, eds. 1991. s.v. "Neoliberalism." In *A Dictionary of Conservative and Libertarian Thought.* London: Routledge.

Aune, James Arnt. 2001. *Selling the free market: The rhetoric of economic correctness.* New York: The Guilford Press.

Barabasi, Albert-Laszlo. 2003. *Linked: How everything is connected to everything else and what it means.* New York: Penguin.

Barbrook, Richard, and Andy Cameron. 1996. The Californian ideology. *Science as Culture* 26:44–72.

Barker, Chris. 2003. *Cultural studies: Theory and practice.* Thousand Oaks, CA: Sage.

Barney, Darin. 2000. *Prometheus wired: The hope for democracy in the age of network technology.* Chicago, IL: The University of Chicago Press.

———. 2004. *The network society.* Cambridge: Polity Press.

Barthes, Roland. 1972. *Mythologies.* New York: Noonday Press.

———. 1982. *Empire of signs.* New York: Hill and Wang.

Battelle, John. 2005. *The search: How Google and its rivals rewrote the rules of business and transformed our culture.* New York: Penguin.

Baudrillard, Jean. 1975. *The mirror of production*. St. Louis, MO: Telos.

———. 1981. *For a critique of the political economy of the sign*. St. Louis, MO: Telos.

———. 1983. *Simulation*. New York: Semiotext(e).

Bauman, Zygmunt. 1998. *Globalization: The human consequences*. New York: Columbia University Press.

———. 2000. *Liquid modernity*. Cambridge: Polity Press.

———. 2001. *The individualized society*. Cambridge: Polity Press.

BBC. 2006a. Interview with Guy Kewney. *News 24*, May 8.

BBC. 2006b. Apology for wrong interviewee. *News Watch*, May 13.

Beck, Ulrich. 1992. *Risk society: Towards a new modernity*. London: Sage.

———. 2000. *The brave new world of work*. Cambridge: Polity Press.

Beck, Ulrich, Anthony Giddens, and Scott Lash. 1994. *Reflexive modernization: Politics, tradition and aesthetics in the modern social order*. Stanford, CA: Stanford University Press.

Bell, Daniel. 1999. *The coming of post-industrial society: A venture in social forecasting*. New York: Basic Books. (Orig. pub. 1973.)

———. 1976. *The cultural contradictions of capitalism*. New York: Basic Books.

Beniger, James. 1986. *The control revolution: Technological and economic origins of the information society*. Cambridge, MA: Harvard University Press.

Benkler, Yochai. 2006. *The wealth of networks: How social production transforms markets and freedom* New Haven, CT: Yale University Press.

Best, Steve, and Douglas Kellner. 2000. Kevin Kelly's complexity theory: The politics and ideology of self-organizing systems. *Democracy and Nature* 6 (3): 375–400.

Bijker, Wiebe. 1995. *Of bicycles, bakelites, and bulbs: Toward a theory of sociotechnical change*. Cambridge, MA: MIT Press.

Bijker, Wiebe, Thomas P. Hughes, and Trevor Pinch, eds. 1989. *The social construction of technological systems: New directions in the sociology and history of technology*. Cambridge, MA: MIT Press.

Bodow, Steve. 2002. Found: Artifacts from the future. *Wired*, June. http://1.bp .blogspot.com/_mEUle6uwKAs/SMbn1L0sSGI/AAAAAAAAA_A/NI3egEIIDEU/ s400/found-06-02-w.jpg.

Boltanski, Luc, and Ève Chiapello. 2005. *The new spirit of capitalism*. London: Verso.

Boltanski, Luc, and Laurent Thevenot. 2006. *On justification: Economies of worth*. Princeton, NJ: Princeton University Press.

Bonabeau, Eric, Marco Dorigo, and Guy Theraulaz. 1999. *Swarm intelligence: From natural to artificial systems*. New York: Oxford University Press.

Borgmann, Albert. 1984a. *Technology and the character of contemporary life*. Chicago, IL: University of Chicago Press.

———. 1984b. Technology and democracy. *Research in Philosophy and Technology*, 7:211–28.

Borsook, Paulina. 1993. Release. *Wired*, November.

———. 1994. Listening to silicon. *Wired*, March.

———. 1996. The anarchist. *Wired*, April.

———. 2000. *Cyberselfish: A critical romp through the terribly libertarian culture of high tech*. New York: PublicAffairs.

Bourdieu, Pierre. 1998a. The essence of neoliberalism. *Le Monde Diplomatique*, December.

———. 1998b. A reasoned utopia and economic fatalism. *New Left Review*, January/February, 125–30.

———. 2003. *Outline of a theory of practice*. Cambridge: Cambridge University Press.

Boyer, Robert. 2002. *Regulation theory: The state of the art*. New York: Routledge.

Brand, Stewart. 1998. Freeman Dyson's brain. *Wired*, February.

Branwyn, Gareth. 1993. The desire to be wired. *Wired*, September/October.

Braverman, Harry. 1974. *Labor and monopoly capital: The degradation of work in the twentieth century*. New York: Monthly Review Press.

Brate, Adam. 2002. *Technomanifestos: Visions from the information revolutionaries*. New York: Texere.

Bronson, Po. 2003. Boom space: What's left after the thrill is gone? *Wired*, June.

Brown, Janelle. 2004. Time warp *Wired*, May.

Browning, John, and Spencer Reiss. 1998. Encyclopedia of the new economy. *Wired*, March/April/May.

Buchanan, Mark, 2002. *Nexus: Small worlds and the groundbreaking science of networks*. New York: Norton.

Callinicos, Alex. 2006. *The resources of critique*. Cambridge: Polity Press.

Callon, Michel. 1991. Techno-economic networks and irreversibility. In *A sociology of monsters: Essays on power, technology and domination*, ed. John Law, 132–61. New York: Routledge.

Capps, Robert. 2004. The humanoid race. *Wired*, July.

Castells, Manuel. 1996. *The rise of the network society*. Vol. 1 of *The information age: Economy, society and culture*. Oxford: Blackwell.

———. 1997. *The power of identity*. Vol. 2 of *The information age: Economy, society and culture*. Oxford: Blackwell.

———. 1998. *End of millennium*. Vol. 3 of *The information age: Economy, society and culture*. Oxford: Blackwell.

Channell, David. 2004. The computer at nature's core. *Wired*, February.

Chaplin, Charles. 1936. *Modern times*. Los Angeles, CA: United Artists.

Chiapello, Eve, and Norman Fairclough. 2002. Understanding the new management ideology: A transdisciplinary contribution from critical discourse analysis and new sociology of capitalism. *Discourse and Society* 13 (2): 185–208.

Chouliaraki, Lilie, and Norman Fairclough. 1999. *Discourse in late modernity: Rethinking critical discourse analysis*. Edinburgh: Edinburgh University Press.

Cooper, Simon. 2002. *Technoculture and critical theory: In the service of the machine?* London: Routledge.

Coupland, Douglas. 1991. *Generation X: Tales for an accelerated culture*. New York: St. Martin's Press.

———. 1994. Microserfs. *Wired*, January.

———. 1995. *Microserfs*. New York: HarperCollins.

Cover. 2000. *Wired*, September.

———. 2002. *Wired*, December.

———. 2003. *Wired*, April.

————. 2004. *Wired*, August.

————. 2005. *Wired*, August.

————. 2006. *Wired*, June.

Cowan, Ruth Schwartz. 1985. *More work for mother: The ironies of household technology from the open hearth to the microwave*. New York: Basic Books.

Coyle, Diane. 1999. *The weightless world: Thriving in the digital age*. Cambridge, MA: MIT Press.

Crumlish, Christian. 2004. *The power of many: How the living web is transforming politics, business, and everyday life*. San Francisco: Sybex.

D'Aluisio, Faith. 2000. At home with the androids. *Wired*, September.

Darrell, Emily. 2007. *Wired* magazine and the evolution of journalism. *New West Missoula*, November 6. http://www.newwest.net/city/article/wired_magazine_and_the_evolution _of_journalism/C8/L8/.

Davis, Erik. 1998. Amiable azoicist. *Wired*, December.

Dawkins, Richard. 1976. *The selfish gene*. New York: Oxford University Press.

Dean, Jodi. 2002. *Publicity's secret: How technoculture capitalizes on democracy*. Ithaca, NY: Cornell University Press.

Dorigo, Marco. 2004. The swarmbots are coming. *Wired*, February.

Duff, Alistair. 2000. *Information society studies*. London: Routledge.

Duggan, Lisa. 2003. *The twilight of equality? Neoliberalism, cultural politics, and the attack on democracy*. Boston: Beacon Press.

Dyer-Witheford, Nick, 1999. *Cyber-Marx: Cycles and circuits of struggle in high-technology capitalism*. Chicago, IL: University of Illinois Press.

Dyson, Esther. 1995. Friend and foe. *Wired*, August.

The Editors of Perseus Publishing. 2002. *We've got blog: How weblogs are changing our culture*. Cambridge, MA: Perseus Publishing.

Ellul, Jacques. 1964. *The technological society*. New York: Knopf.

Feenberg, Andrew. 1991. *Critical theory of technology*. New York: Oxford University Press.

————. 1995. Subversive rationalization: Technology, power, and democracy. In *Technology and the Politics of Knowledge*, ed. Andrew Feenberg and Alastair Hannay, 3–22. Indianapolis, IN: Indiana University Press.

————. 1996. Marcuse or Habermas: Two critiques of technology. *Inquiry* 39:45–70.

Fisher, Eran. 2007. From safety net to the Internet: The discourse on network production in post-Fordist society. In *New media and innovative technologies*, ed. Tal Azran and Dan Caspi, 98–131. Be'er Sheva: Ben-Gurion University Press.

Florida, Richard. 2003. *The rise of the creative class: And how it's transforming work, leisure, community and everyday life*. New York: Basic Books.

Foucault, Michel. 1994. *The order of things: An archeology of human sciences*. New York: Vintage.

————. 1995. *Discipline and punish: The birth of the prison*. New York: Vintage.

————. 2002. *Archeology of knowledge*. New York: Routledge.

Frank, Thomas. 2000. *One market under God: Extreme capitalism, market populism, and the end of economic democracy*. New York: Anchor Books.

Frank, Thomas, and Matt Weiland, eds. 1997. *Commodity your dissent: Salvos from the baffler*. New York: W. W. Norton & Company.

Fraser, Nancy. 2003. From discipline to flexibilization? Rereading Foucault in the shadow of globalization. *Constellations* 10 (2): 160–71.

Fraser, Nancy, and Axel Honneth. 2003. *Redistribution of recognition? A political-philosophical exchange*. New York: Verso.

Freud, Sigmund. 1963. *Jokes and their relation to the unconscious*. New York: W. W. Norton & Company.

Freund, Jesse. 2004. Bionic ears. *Wired*, May.

Friedman, Milton. 2006. Free markets and the end of history. *New Perspectives Quarterly* 23 (1). http://www.digitalnpq/archive/2006_Winter/friedman.html.

Friedman, Thomas. 2000. *The Lexus and the olive tree: Understanding globalization*. New York: Anchor Books.

Frith, Lucy. 1994. *Society, dichotomies and resolutions: An inquiry into social synthesis*. Hampshire: Ashgate Publishing.

Fuchs, Christian. 2008. *Internet and society: Social theory in the information age*. New York: Routledge.

Fukuyama, Francis. 1992. *The end of history and the last man*. New York: Free Press.

Fuller, Steve. 1999. Creative destruction. In *The New Fontana Dictionary of Modern Thought*, ed. Alan Bullock and Stephen Trombley, 180–81. New York: Harper Collins.

Gaggi, Silvio. 2003. The cyborg and the net: Figures of the technological subjects. In *Adrift in the technological matrix*, ed. David Erben, 125–39. Lewisburg, PA: Bucknell University Press.

Gamble, Andrew. 1996. *Hayek: The iron cage of liberty*. Boulder, CO: Westview.

Geertz, Clifford. 1977. *The interpretation of cultures*. New York: Basic Books.

Gere, Charlie. 2002. *Digital culture*. London: Reaktion Books.

Gilder, George. 1990. *Microcosm: The quantum revolution in economics and technology*. New York: Free Press.

———. 1998. Happy birthday wired. *Wired*, January.

Glenny, Misha. 2001. How Europe can stop worrying and learn to love the future. *Wired*, February.

Glickman, Adam. 1999. Softwear. *Wired*, December.

Goetz, Thomas. 2003. Open source everywhere. *Wired*, November.

Goggins, William. 1998. Media odyssey. *Wired*, March.

Gramsci, Antonio. 1971. Americanism and Fordism. In *Selections from the Prison Notebooks*, trans. and ed. Quintin Hoare and Geoffrey Nowell-Smith, 277–318. New York: International Publishers.

Granovetter, Mark. 1973. The strength of weak ties, *American Journal of Sociology*, 78: 1360–80.

Greenbaum, Joan. 1995. *Windows on the workplace: Computers, jobs, and the organization of office work in the late twentieth century*. New York: Monthly Review Books.

Greenwald, Douglas, ed. 1994. Liberalism. In *The McGraw-Hill encyclopedia of economics*. 2nd ed. New York: McGraw-Hill.

Habermas, Jürgen. 1970. Technology and science as "ideology." In *Toward a rational society: Student protest, science, and politics*, 81–122. Boston: Beacon Press.

———. 1973. *Legitimation crisis*. Boston: Beacon Press.

———. 1984. *Lifeworld and system: A critique of functionalist reason*. Vol. 2 of *The theory of communicative action*. Boston: Beacon Press.

Hafner, Katie. 1997. The epic saga of the WELL. *Wired*, May.

Hamel, Jacques, Stephane Dufour, and Dominic Fortin. 1991. *Case study methods*. Newbury Park, CA: Sage.

Hård, Mikael. 1994. *Machines are frozen spirit: The scientification of refrigeration and brewing in the 19th century—a Weberian interpretation*. Boulder, CO: Westview Press.

Hardt, Michael, and Antonio Negri. 2001. *Empire*. Cambridge, MA: Harvard University Press.

———. 2005. *Multitude: War and democracy in the age of empire*. New York: Penguin.

Haraway, Donna. 1991. *Simians, cyborgs, and women: The reinvention of nature*. New York: Routledge.

———. 1997. *ModestWitness@SecondMillenium. FemaleMan meets OncoMouse: Feminism and technoscience*. New York: Routledge.

Harrison, Bennett. 1997. *Lean and mean: Why large corporations will continue to dominate the global economy*. New York: Guilford Press.

Harvey, David. 1989. *The condition of postmodernity: An enquiry into the origins of cultural change*. Oxford: Blackwell.

———. 2005. *A brief history of neoliberalism*. Oxford: Oxford University Press.

———. 2009. *Cosmopolitanism and the geographies of freedom*. New York: Columbia University Press.

Hayek, Friedrich. 1967. *Studies in philosophy, politics and economics*. Chicago: University of Chicago Press.

———. 1979. *The counter-revolution of science: Studies on the abuse of reason*. Indianapolis: Liberty Fund.

———. 1981. *The political order of a free people*. Vol. 3 of *Law, legislation and liberty*. Chicago: Chicago University Press.

Hayles, M. Katherine. 1997. The posthuman body: Inscription and incorporation in "Galatea 2.2" and "Snow Crash." *Configurations* 5 (2): 241–66.

———. 1999. *How we became posthuman: Virtual bodies, cybernetics, literature, and informatics*. Chicago, IL: The University of Chicago Press.

Head, Simon. 2003. *The new ruthless economy: Work and power in the digital age*. Oxford: Oxford University Press.

Heffernan, Nick. 2000. *Capital, class, and technology in contemporary American culture: Projecting post-Fordism*. London: Pluto Press.

Heidegger, Martin. 1977. The question concerning technology. In *The question concerning technology and other essays*, 3–35. New York: Harper Torchbooks.

Heilemann, John. 2001. Andy Grove's rational exuberance. *Wired*, June.

Herf, Jeffrey. 1984. *Reactionary modernism: Technology, culture, and politics in Weimar and the Third Reich*. Cambridge: Cambridge University Press.

Hillis, Danny. 1998. The big picture. *Wired*, January.

Holland, John. 1996. *Hidden order: How adaptation builds complexity.* New York: Basic Books.

Horkheimer, Max, and Theodor Adorno. 1976. *Dialectics of enlightenment.* New York: Continuum.

Howe, Jeff. 2003. The connectors. *Wired,* November.

———. 2006. The rise of crowdsourcing. *Wired,* June.

Hughes, Dave. 1994. Chaos is the form. *Wired,* January.

Hunt, Lynn. 1984. *Politics, culture and class in the French Revolution.* Berkeley, CA: University of California Press.

Huws, Ursula. 2003. *The making of a cybertariat: Virtual work in real world.* New York: Monthly Review Press.

Jameson, Fredric. 1982. *The political unconscious: Narratives as socially symbolic act.* Ithaca, NY: Cornell University Press.

———. 1991. *Postmodernism, or The cultural logic of late capitalism.* London: Verso.

Jessop, Bob. 1994. Post-Fordism and the state. In *Post-Fordism,* ed. Ash Amin, 251–79. Oxford: Blackwell.

Johnson, Steven. 2002. *Emergence: The connected lives of ants, brains, cities, and software.* New York: Touchstone.

Jonscher, Charles. 1999. *The evolution of wired life: From the alphabet to the soul-catcher chip—how information technologies change our world.* Hoboken, NJ: Wiley.

Kahn, Jennifer. 2003. Regrow your own. *Wired,* November.

Kaplan, Seth. 2005. Found: Artifact from the future. *Wired,* July. http://www.wired .com/wired/archive/13.07/images/found.jpg.

Katz, Jon. 1995. Lost world of the future. *Wired,* October.

Kelly, Kevin. 1993. George Gilder: When bandwidth is free. *Wired,* September/October.

———. 1995. Singular visionary. *Wired,* June.

———. 1997. New rules for the new economy. *Wired,* September.

———. 1998. *New rules for the new economy: 10 radical strategies for a connected world.* New York: Viking.

———. 1999. The roaring zeros. *Wired,* September.

———. 2002. God is the machine. *Wired,* December. http://www.wired.com/wired/ archive/10.12/holytech.html.

———. 2005. We are the web. *Wired,* August.

Kennedy, Lisa. 2002. Spielberg in the twilight zone. *Wired,* June.

Kern, Stephen. 1983. *The culture of time and space, 1880–1918.* Cambridge, MA: Harvard University Press.

Kessler, Andy. 2006. *The end of medicine: How Silicon Valley (and naked mice) will reboot tour doctor.* New York: Collins.

Klein, Naomi. 1999. *No logo: Taking aim at the brand bullies.* New York: Picador.

Kley, Roland. 1994. *Hayek's social and political thought.* Oxford: Oxford University Press.

Kline, David, and Dan Burstein. 2005. *Blog!: How the newest media revolution is changing politics, business, and culture.* New York: CDS Books.

Kodak. 1999. Advertisement. *Wired,* December.

Koerner, Brendan, I. 2006. Geeks in toyland. *Wired*, February.

Koolhaas, Rem. 2003. Migrant labor: Office space: Where do you want to work today? *Wired*, June.

Kunda, Gideon. 1992. *Engineering culture: Control and commitment in a high-tech corporation*. Philadelphia: Temple University Press.

Kunzru, Hari. 1997. You are cyborg. *Wired*, February.

Kuper, Adam, and Jessica Kuper, eds. 1996. *The social science encyclopedia*. London: Routledge.

Kurzweil, Ray. 2000. *The age of spiritual machines: When computers exceed human intelligence*. New York: Viking.

———. 2005. *The singularity is near: When humans transcend biology*. New York: Viking.

Lang, Stacey Smith. 2001. Evolution. *Wired*, October.

Lash, Scott. 2002. *Critique of information*. London: Sage.

Lash, Scott, and John Urry. 1987. *The end of organized capitalism*. Oxford: Polity Press.

Latour, Bruno. 1991. Technology is society made durable. In *A sociology of monsters: Essays on power, technology, and domination*, ed. John Law, 103–30. London: Routledge.

———. 2005. *Reassembling the social: An introduction to actor-network-theory*. Oxford: Oxford University Press.

Law, John, and John Hassard, eds. 1999. *Actor network theory and after*. Boston: Blackwell Publishers.

Lévi-Strauss, Claude. 1963. *Structural anthropology*. New York: Anchor Books.

Levinson, Paul. 1998. *The soft edge: A natural history and future of the information revolution*. New York: Routledge.

Levy, Pierre. 1997. *Collective intelligence: Mankind's emerging world in cyberspace*. Boulder, CO: Perseus Books.

Levy, Steven. 2005. The trend spotter. *Wired*, October.

Liu, Alan. 2004. *The laws of cool: Knowledge work and the culture of information*. Chicago, IL: The University of Chicago Press.

Lloyd, Richard. 2005. *Neo-bohemia: Art and commerce in the postindustrial city*. New York: Routledge.

Lovink, Geert. 2002. *Uncanny networks: Dialogues with the virtual intelligentsia*. Cambridge, MA: MIT Press.

Lowe, Donald. 1995. *The body in late-capitalist USA*. Durham, NC: Duke University Press.

Luman, Stuart. 2005. Open source softwear. *Wired*, June.

Lutz, Catherine, and Jane Collins. 1993. *Reading National Geographic*. Chicago, IL: The University of Chicago Press.

Lyotard, Jean-François. 1984. *The postmodern condition: A report on knowledge*. Minneapolis, MN: University of Minnesota Press.

Machlup, Fritz. 1962. *The production and distribution of knowledge in the United States*. Princeton, NJ: Princeton University Press.

Mackay, Hugh. 2003. *Investigating the information society*. London: Routledge.

MacKenzie, Donald. 1996. *Knowing machines: Essays on technological change*. Cambridge, MA: MIT Press.

MacKenzie, Donald, and Judy Wajcman, eds. 1999. *The social shaping of technology*. rev. 2nd ed. Philadelphia: Open University Press.

Marcuse, Herbert. 1974. *Eros and civilization: A philosophical inquiry into Freud*. Boston: Beacon Press.

———. 1991. *One dimensional man: Studies in the ideology of advanced industrial society*. 2nd ed. Boston: Beacon Press.

Marshall, Gordon, ed. 1994. *Concise dictionary of sociology*. Oxford: Oxford University Press.

Martin, Richard. 2005. Mind control. *Wired*, March.

Marx, Karl. 1990. *Capital*, Vol. 1. New York: Peguin Books. (Orig. pub. 1867.)

———. 1995. *The poverty of philosophy*. Amherst, MA: Prometheus Books. (Orig. pub. 1847.)

Mattelart, Armand. 2003. *The information society: An introduction*. Thousand Oaks, CA: Sage.

May, Christopher. 2002. *The information society: A skeptical view*. Cambridge: Polity Press.

———. 2003. *Key thinkers for the information society*. London: Routledge.

Mayr, Otto. 1986. *Authority, liberty, and automatic machinery in early modern Europe*. Baltimore, MD: Johns Hopkins University Press.

McCorduck, Pamela. 1996. Sex, lies and avatars. *Wired*, April.

Meyer, Christopher. 2004. The new facts of life. *Wired*, February.

Milanovic, Branko. 2007. *Worlds apart: Measuring international and global inequality*. Princeton, NJ: Princeton University Press.

Mitchell, William J. 2004. *Me++: The cyborg self and the networked city*. Cambridge, MA: MIT Press.

Moravec, Hans. 1999. *Robot: Mere machine to transcendent mind*. New York: Oxford University Press.

Moreno, Shonquis. 2002. Second skin. *Wired*, May.

Mosco, Vincent. 2004. *The digital sublime: Myth, power, and cyberspace*. Cambridge, MA: MIT Press.

Murphie, Andrew, and John Potts. 2003. *Culture and technology*. New York: Palgrave Macmillan.

Negroponte, Nicholas. 1996. *Being digital*. New York: Vintage.

———. 1997. Negroponte. *Wired*, October.

Noble, David. 1984. *Forces of production: A social history of industrial automation*. New York: Knopf.

———. 1995. *Progress without people: New technology, unemployment, and the message of resistance*. Toronto: Between the Lines.

———. 1999. *The religion of technology: The divinity of man and the spirit of invention*. New York: Penguin Books.

Nye, David. 1994. *American technological sublime*. Cambridge, MA: MIT Press.

———. 2003. *America as second creation: Technology and narratives of new beginnings*. Cambridge, MA: MIT Press.

Offe, Claus. 1984a. *Contradictions of the welfare state*. Cambridge, MA: MIT Press.

———. 1984b. *Disorganized capitalism: Contemporary transformations of work and politics.* Cambridge, MA: The MIT Press.

Olcese, James. 2004. Measuring up. *Wired*, September.

Olympus. 1999. Advertisement. *Wired*, December.

Oram, Andy. 2001. *Peer-to-peer: Harnessing the power of disruptive technologies.* Cambridge, MA: O'Reilly.

Perelman, Michael. 2006. *Railroading economics: The creation of the free market mythology.* New York: Monthly Review Press.

Peters, Michael. 2001. *Poststructuralism, Marxism, and neoliberalism: Between theory and politics.* New York: Rowman & Littlefield.

Petsoulas, Christina. 2001. *Hayek's liberalism and its origins: His idea of spontaneous order and the Scottish enlightenment.* New York: Routledge.

Pink, Daniel. 2005. Revenge of the right brain. *Wired*, February. http://www.wired.com/wired/images.html?issue=13.02&topic=brain&img=1.

———. 2006. *A whole new mind: Why right-brainers will rule the future.* New York: Penguin.

Pippin, Robert. 1995. On the notion of technology as ideology. In *Technology and the Politics of Knowledge*, ed. Andrew Feenberg and Alastair Hannay, 43–61. Indianapolis, IN: Indiana University Press.

Piven, Frances Fox, and Richard Cloward. 1997. *The breaking of the American social compact.* New York: The New Press.

Platt, Charles. 1995a. What's it mean to be human, anyway? *Wired*, April.

———. 1995b. Superhumanism. Wired, October.

———. 2000. Steaming video. *Wired*, November.

Polanyi, Karl. 2001. *The great transformation: The political and economic origins of our time.* Boston: Beacon Press. (Orig. pub. 1944.)

Porat, Marc, and Michael Rubin. 1977. *The information economy.* 9 vols. Washington, DC: U.S. Department of Commerce, Office of Telecommunications.

Poster, Mark. 1990. *The mode of information: Poststructuralism and social context.* Cambridge: Polity Press.

Postman, Neil. 1993. *Technopoly: The surrender of culture to technology.* New York: Vintage Books.

Postrel, Virginia. 1998. Technocracy R.I.P. *Wired*, January.

Rabinbach, Anson. 1992. *The human motor: Energy, fatigue, and the origins of modernity.* Berkeley, CA: University of California Press.

Ram, Uri. 2006. *The time of the post: Nationalism and the politics of knowledge in Israel.* Tel-Aviv: Resling.

———. 2007. *The globalization of Israel: McWorld in Tel-Aviv, jihad in Jerusalem.* New York: Routledge.

Rapley, John. 2004. *Globalization and inequality: Neoliberalism's downward spiral.* Boulder, CO: Lynne Rienner Publishers.

Regis, Ed. 1994. Meet the extropians. *Wired*, October.

Reich, Robert. 1991. *The work of nations: Preparing ourselves for 21st century capitalism.* New York: Vintage.

Reynolds, Glenn. 2006. *An army of Davids: How markets and technology empower ordinary people to beat big media, big government, and other Goliaths.* Nashville: Thomas Nelson.

Rheingold, Howard. 2003. *Smart mobs: The next social revolution.* New York: Basic Books.

Rivlin, Gary. 2003. Leader of the free world. *Wired,* November.

Robins, Kevin, and Frank Webster. 1985. *Information technology: A Luddite analysis.* Norwood, NJ: Ablex.

———. 1999. *Times of technoculture: From the information society to the virtual life.* London: Routledge.

Rohm, Wendy Goldman. 2004. Seven days of creation. *Wired,* January.

Rosa, Hartmut. 2003. Social acceleration: Ethical and political consequences of a desynchronized high-speed society. *Constellations* 10 (1): 3–33.

Ross, Andrew. 2003. *No-collar: The humane workplace and its hidden costs.* New York: Basic Books.

Rossetto, Louis. 1998. Change is good. *Wired,* January.

Said, Edward. 1981. *Covering Islam: How the media and the experts determine how we see the rest of the world.* New York: Pantheon Books.

Sally, Razeen. 1998. *Classical liberalism and the international economic order: Studies in theory and intellectual history.* London: Routledge.

Sassen, Saskia. 1999. *Globalization and its discontents: Essays on the new mobility of people and money.* New York: New Press.

Schrage, Michael. 1994. Is advertising dead? *Wired,* February.

———. 1995. Revolutionary evolutionist. *Wired,* July.

Schwandt, Thomas. 2001. *Dictionary of qualitative inquiry.* Thousand Oaks, CA: Sage.

Schwartz, Peter. 1993. Post-capitalist. *Wired,* July/August.

Schwartz, Peter, and Kevin Kelly. 1996. The relentless contrarian. *Wired,* August.

Scoble, Robert, and Shel Israel. 2006. *Naked conversations: How blogs are changing the way businesses talk with customers.* Hoboken, NJ: Wiley.

Scott, Joan. 1988. Deconstructing equality-versus-difference: Or, the uses of post-structuralist theory for feminism. *Feminist Studies* 14 (1): 33–50.

Segal, Howard. 1985. *Technological utopianism in American culture.* Chicago, IL: The University of Chicago Press.

———. 1994. *Future imperfect: The mixed blessings of technology in America.* Amherst, MA: The University of Massachusetts Press.

Sellers, Michael. 1995. Memetic correction. *Wired,* January.

Sennet, Richard. 2000. *The Corrosion of character: The personal consequences of work in the new capitalism.* New York: Norton.

———. 2006. *The culture of the new capitalism.* New Haven, CT: Yale University Press.

Shannon, Claude, and Warren Weaver. 1998. *The mathematical theory of communication.* Chicago, IL: University of Illinois Press.

Shilling, Chris. 2005. *The body in culture, technology and society.* London: Sage.

Silberman, Steve. 2006. Don't even think about lying. *Wired,* January. http://www.wired.com/wired/images.html?issue=14.01&topic=lying&img=2.

Simpson, Roderick. 1997. Cloning. Problem? No problem. *Wired*, September.

Sklair, Leslie. 2002. *Globalization: Capitalism and its alternatives*. Oxford: Oxford University Press.

Smith, Adam. 1982. *The wealth of nations*. New York: Penguin Books. (Orig. pub. 1776).

Smith, Merritt, and Leo Marx, eds. 1994. *Does technology drive history? The dilemma of technological determinism*. Cambridge, MA: MIT Press.

Somers, Margaret, and Fred Block. 2005. From poverty to perversity: Ideas, markets and institutions over 200 years of welfare debate. *American Sociological Review* 70:260–87.

Sony. 1999a. Advertisement. *Wired*, December.

———. 1999b. Advertisement. *Wired*, December.

Stake, Robert. 1994. Case studies. In *Handbook of Qualitative Research*, ed. Norman K. Denzin and Yvonna S. Lincoln, 236–47. Thousand Oaks, CA: Sage.

Stehr, Nico. 2001. *The fragility of modern societies: Knowledge and risk in the information age*. Thousand Oaks, CA: Sage.

Steichen, Edward. 2002. *The family of man*. New York: Museum of Modern Art. (Orig. pub. 1955.)

Sterling, Bruce. 2005. "Order out of chaos, *Wired*, April.

Stevenson, Nick, Peter Jackson, and Kate Brooks. 2001. *Making sense of men's magazines*. Cambridge: Polity Press.

Stewart Millar, Melanie. 1996. Net (e)scape: A feminist analysis of the digital discourse of *Wired* magazine. MA thesis, York University.

Stone, Biz. 2004. *Who let The blogs out?: A hyperconnected peek at The world of weblogs*. New York: St. Martin's Press.

Strogatz, Steven. 2003. *Sync: The emerging science of spontaneous order*. New York: Hyperion.

Sturken, Marita, and Douglas Thomas. 2004. Introduction: Technological visions and the rhetoric of the new. In *Technological visions: The hopes and fears that shape new technologies*, ed Marita Sturken, Douglas Thomas, and Sandra J. Bell-Rokeach, 1–18. Philadelphia: Temple University Press.

Surowiecki, James. 2004. *The wisdom of crowds: Why the many are smarter than the few and how collective wisdom shapes business, economics, societies and nations*. New York: Doubleday.

T-Mobile. Advertisement, *Wired*, October 2004.

Tanz, Jason. 2007. Desktop R.I.P. *Wired*, March.

Taylor, Frederick Winslow. 1967. *Principles of scientific management*. New York: Norton.

Taylor, Mark C. 2003. *The moment of complexity: Emerging network culture*. Chicago, IL: University of Chicago Press.

Toffler, Alvin. 1980. *The third wave*. New York: Bantam Books.

Touraine, Alain. 1971. *The post-industrial society: Tomorrow's social history: Classes, conflict and culture in the programmed society*. New York: Random House.

Turkle, Sherry. 1996. Who Am We? *Wired*, January.

———. 1997. *Life on the screen: Identity in the age of the internet*. New York: Simon & Schuster.

Turner, Fred. 2006. *From counterculture to cyberculture: Stewart Brand, the whole earth network, and the rise of digital utopianism.* Chicago, IL: University of Chicago Press.

Uchitelle, Louis. 1994. Insecurity forever: The rise of the losing class. *New York Times,* November 20.

Urry, John. 2003. *Global complexity.* Cambridge: Polity Press.

Verizon Wireless. 2004. Advertisement. *Wired,* August.

Von Hippel, Eric. 2006. *Democratizing innovation.* Cambridge, MA: MIT Press.

Wajcman, Judy. 2004. *Technofeminism.* Cambridge: Polity Press.

Wallerstein, Immanuel. 2004. *World-system analysis: An introduction.* Durham, NC: Duke University Press

Weber, Max. 1921. *Economy and society: An outline of interpretive sociology,* ed. Guenther Roth and Claus Wittich. Berkeley: University of California Press, 1978.

———. 1958. *The Protestant ethic and the spirit of capitalism.* New York: Scribners.

Webster, Andrew. 1991. *Science, technology, and society: New directions.* New Brunswick, NJ: Rutgers University Press.

Webster, Frank. 2002. *Theories of the information society.* 2nd ed. London: Routledge.

———. 2005. Making sense of the information age: Sociology and cultural studies. *Information, Communication & Society* 8 (4): 439–58.

Westbury, Chris. 1995. Propagating the meme meme. *Wired,* April.

Wiener, Norbert. 1965. *Cybernetics: The control and communication in the animal and the machine.* Cambridge, MA: MIT Press.

Williams, Raymond. 1978. *Marxism and literature.* New York: Oxford University Press.

Williams, Robin, and David Edge. 1996. The social shaping of technology. *Research Policy* 25:856–99.

Winner, Langdon. 1977. *Autonomous technology: Technics-out-of-control as a theme in political thought.* Cambridge, MA: MIT Press.

Wired editors. 1998. Meme. *Wired,* February.

———. 2004a. Living machines. *Wired,* February.

———. 2004b. Evolve. *Wired,* February.

Wolf, Gary. 2003. *Wired: A romance.* New York: Random House.

Wriston, Walter. 1992. *The twilight of sovereignty: How the information revolution is transforming our world.* New York: Scribner.

Yergin, Daniel, and Joseph Stanislaw. 1998. *The commanding heights: The battle between government and the marketplace that is remaking the modern world.* New York: Simon and Schuster.

Yin, Robert K. 1989. *Case study research: Design and method.* Newbury Park, CA: Sage.

Žižek, Slavoj. 2005. *Iraq: The borrowed kettle.* New York: Verso.

INDEX

GPSR Compliance
The European Union's (EU) General Product Safety Regulation (GPSR) is a set
of rules that requires consumer products to be safe and our obligations to
ensure this.

If you have any concerns about our products, you can contact us on

ProductSafety@springernature.com

In case Publisher is established outside the EU, the EU authorized
representative is:

Springer Nature Customer Service Center GmbH
Europaplatz 3
69115 Heidelberg, Germany